The Effects of Greenhouse Gas Reduction Policy on Global Economy

Zhengning Pu

Science Press
Beijing

Responsible Editors: Kai Hu, Lei Xu

Copyright © 2017 by Science Press
Published by Science Press
16 Donghuangchenggen North Street
Beijing 100717, P. R. China

Printed in Beijing

All rights reserved. No part of this publication may be reproduced, stored in a retrieval system, or transmitted in any form or by any means, electronic, mechanical, photocopying, recording or otherwise, without the prior written permission of the copyright owner.

ISBN 978-7-03-054165-9

Preface

1. Introduction

Facing more and more extreme environmental events, the whole human society has realized that more attention should be paid to protect this planet's environment. As a result, the United Nations Framework Convention on Climate Change (UNFCCC) had been settled as the first worldwide formal framework to settle problems brought by climate changes. But two decades after the UNFCCC was constructed, the proposition on reducing global greenhouse gas (GHG) is still in a long game.

In this research, a Multi-Regional Computable General Equilibrium (MRCGE) was created to evaluate the possible energy resource tax's effect for different regions of China. Furthermore, such a model could form a linkage to the Global Trade Analysis Project (GTAP) model as a new analysis tool, which considered both the inner country regional division of China and world region division to evaluate the carbon tax effectiveness in China and worldwide.

2. MRCGE Analysis for China's Energy Resource Tax Policy

Scenarios

Scenarios for this analysis were considered from two different angles. On the one hand, because China wanted to change their energy resource policy from almost zero tax to a 5% *ad valorem* tax on all energy goods for controlling carbon emission and because the policy would have the first pilot in the western area of China, we decided to observe the possible effect caused by this pilot policy and the effects of future policies that might be followed. On the other hand, since China promised a 40%~45%

per GDP CO_2 emission reduction during 2009 Copenhagen meeting, we investigated the policy impacts of this emissions reduction commitment.

To achieve these goals, we followed China's 2011 western area test fiscal policy (5% energy tax for energy resources) and its promised aim, setting up six different scenarios to evaluate the effectiveness of China's energy tax policy.

Simulation results

With the scenario simulation, two major conclusions were found in this MRCGE analysis. First, a significant CO_2 reduction could be accomplished under the performance of energy resource tax in certain ranges while the minor adverse effects on macroeconomic of China were made.

Second, as the industry structure and economic level of development is different through 8 regions, the execution of energy resource tax may lead to unfairness for regional per capita household utility losses.

3. MRCGE-GTAP Linkage Model Analysis

Scenarios

Five different scenarios were designed to evaluate the effects of execution of carbon tax in industrialized countries, China and the entire world. Goals for scenarios were referred to the Global carbon dioxide emission reduction target mentioned in Kyoto Protocol and the major emitter's commitment on COP15 United Nations Climate Change Conference held in Copenhagen.

Simulation results

Five conclusions can be obtained from the simulation results.

First, execution of carbon tax can effectively reduce carbon dioxide emission worldwide. Furthermore, if main industrialized countries execute the same level of carbon tax, the Kyoto Protocol global target is achievable.

Second, carbon tax may have negative effects on major developed countries, especially on Japan, USA and EU member countries.

Third, worldwide carbon tax use without China may benefit China's household final users by increasing their consumption utility and may have negative effect on China's main export industries.

Fourth, this simulation shows that carbon tax will also cause decrease in household final user's consumption utility in different regions of China, when the nation applies to carbon tax policy.

Finally, as the analysis of world range industry output rate of change shows, no matter developed countries joined only or worldwide in the arrangement of the carbon tax, least developed countries will benefit from such apolicy arrangement.

4. Concluding remarks

In this research, a one country multi-regional CGE model and a MRCGE-GTAP model were constructed to evaluate the effect of China's climate policy. The main conclusions are obtained as follows.

First, China's Copenhagen guarantee can be reached with the base year total CO_2 emission as the exogenous parameter in the model. The effectiveness will decrease when the total CO_2 emission of the nation is treated as endogenous.

Second, an energy resource tax/carbon tax will produce regional differential in household utility losses since there is a gap of economic development between different regions in China. Therefore, the possibility of fairness should be considered in real policy execution.

Third, a worldwide carbon tax levying, especially levying in industrialized nations can effectively reduce the global carbon emission and reach the reduction common goal of the human society which has been written into Kyoto Protocol.

Finally, it is also found that even if the carbon tax is only executed in a few countries besides China, its economy effect will be worldwide including China.

Contents

Preface
1. Origin of Greenhouse Gas Issue 1
2. Literature Review 5
    2.1   What is CGE Model? 5
    2.2   CGE Analysis on Carbon Tax Effect Worldwide 7
    2.3   CGE Analysis on Carbon Tax Effect in China 9
    2.4   Current Multi-Regional CGE Models 11
        2.4.1   MMRF-Green Model 11
        2.4.2   GTAP-E Model 17
3. Responses to Current Climate Change 23
    3.1   International Responses 23
    3.2   Nation and Regional Political Responses 28
    3.3   Responses from Multinational Cooperations 30
    3.4   China's Responses 33
4. Will China's Energy Resource Tax Policy Work? 41
    4.1   Model Structure 41
    4.2   Data and Scenario 48
        4.2.1   Regional Division 48
        4.2.2   Industry Classification 50
        4.2.3   Scenarios 51
    4.3   Analysis Results 52
        4.3.1   National Level Effect 52
        4.3.2   Regional Level Effect 56
        4.3.3   Industry Indicator 60

4.4	A Dynamic Extension		62
	4.4.1	Model Extension	62
	4.4.2	Data Resource and Scenario Setup	63
	4.4.3	Results	64
4.5	Conclusion		69

5. What Will Happen When the World Works Together? ... 71

5.1	Model Structure		71
5.2	Data and Scenario		81
	5.2.1	Regional Division	81
	5.2.2	Industry Classification	82
	5.2.3	Scenarios	84
5.3	Analgsis Results		86
	5.3.1	Changes in EV and CO_2 Emission Reductions in China	86
	5.3.2	China's Industry Changes	88
	5.3.3	Change of Household EV and CO_2 Reduction of World Regions	88
	5.3.4	Industrial Output in World Regions	91
5.4	Conclusion		100

6. Concluding Remarks ... 103

6.1	Simulation Model Construction	103
6.2	Analysis for China	104
6.3	Analysis for the World	106
6.4	Conclusion	106

Reference ... 107

Appendix ... 111

Model 1: One National, Multi-Region, Static Model ... 112
Model 2: Multi-National, Multi-Region, Static Model ... 130
Model 3: One National, Multi-Region, Dynamic Model ... 196

Acknowledgment ... 212

Tables and Charts

Fig. 2.1	Household Demand in MMRF-Green	12
Fig. 2.2	Production Structure in MMRF-Green besides Electricity Sector	13
Fig. 2.3	Production Structure in MMRF-Green Electricity Sector	14
Fig. 2.4	Creation of Capital Goods in MMRF-Green	15
Fig. 2.5	GTAP-E Production Structure	19
Fig. 2.6	GTAP-E Capital-Energy Composite Structure	19
Fig. 2.7	GTAP-E Government Purchases	20
Fig. 2.8	GTAP-E Household Private Purchases	21
Fig. 4.1	Production Structure	42
Fig. 4.2	Household Activities	43
Fig. 4.3	Government Activities	44
Fig. 4.4	Export Structure	45
Fig. 4.5	Import Structure	46
Fig. 4.6	CO_2 Reduction and Energy Intensity Rate of Change	54
Fig. 4.7	National MAC under Different Scenarios	55
Fig. 4.8	Regional GDP Rate of Change	57
Fig. 4.9	Regional Household Utility Change	57
Fig. 4.10	Per Capita Household Utility Change	58
Fig. 4.11	Regional Petroleum and Natural Gas Mining Industry Output Reduction	59
Fig. 4.12	Regional Coal Mining Industry Output Reduction	60
Fig. 4.13	Industry Output Rate of Change	61
Fig. 4.14	Regional Investment Rate of Change	66
Fig. 4.15	Regional Household Consumption Rate of Change	66
Fig. 4.16	Regional GRP Rate of Change	67

Fig. 4.17   National GDP Rate of Change ··· 67
Fig. 4.18   Regional Coal Mining Output Rate of Change ······························· 68
Fig. 4.19   Regional Oil and Gas Mining Output Rate of Change ····················· 69
Fig. 5.1   Production Structure ·· 72
Fig. 5.2   Household Activities ·· 72
Fig. 5.3   Government Activities ··· 73
Fig. 5.4   Export Structure ·· 73
Fig. 5.5   Import Structure ·· 73
Fig. 5.6   Production Structure for World Region ································· 74
Fig. 5.7   Private Sector Activities for World Region ··························· 76
Fig. 5.8   Government Activities for World Region ································ 77
Fig. 5.9   Export Structure for World Region ······································ 78
Fig. 5.10   Import Structure for World Region ······································ 79
Fig. 5.11   China's Regional EV Change ··· 86
Fig. 5.12   China Regional CO_2 Emission Reduction Rate of Change ··············· 87
Fig. 5.13   China's Industry Output Rate of Change ································ 89
Fig. 5.14   World Region EV Change ··· 90
Fig. 5.15   World CO_2 Emission Reduction Rate under Different Scenarios ········ 91
Fig. 5.16   Industry Output Rate of Change of "OCN" Area ·························· 92
Fig. 5.17   Industry Output Rate of Change of "JPN" Area ·························· 93
Fig. 5.18   Industry Output Rate of Change of "GCA" Area ·························· 94
Fig. 5.19   Industry Output Rate of Change of "IND" Area ··························· 95
Fig. 5.20   Industry Output Rate of Change of "ROA" Area ·························· 96
Fig. 5.21   Industry Output Rate of Change of "RUS" Area ··························· 97
Fig. 5.22   Industry Output Rate of Change of "USA" Area ·························· 98
Fig. 5.23   Industry Output Rate of Change of "EU27" Area ························· 99
Fig. 5.24   Industry Output Rate of Change of "ROW" Area ························· 100
Table 3.1   Details about Each COP Conference ······································· 23
Table 4.1   Regional Division Code ·· 48
Table 4.2   Reclassified Commodity Sectors ·· 50
Table 4.3   Scenario Setup ·· 52
Table 4.4   National Level Indexes ·· 53
Table 4.5   Regional Industry Output Rate of Change in the Year 2012 ················ 65

Table 5.1　World's Regional Division ·· 82
Table 5.2　Reclassified Commodity Sectors ·· 83
Table 5.3　Reduction Target for Developed Countries ·························· 85
Table 5.4　Scenario Setup ·· 85
Picture 4.1　Region Code for 8 Division Regions in China ····················· 48
Picture 5.1　Regional Division of Worldwide ······································· 82

Origin of Greenhouse Gas Issue   1

The 200 years since the industrial age began has probably been the most dazzling period in the history of mankind so far. The range of resources, our greatly enhanced ability to harness energy for our own benefit, the wealth created in society as a whole, economic development and advances in science and technology, have reached a point that would astonish people of earlier times. Certainly, this unparalleled growth and development has also brought unparalleled destruction and consumption. Just take modern times as an example: human consumption of primary-class energy rose from about 3,813.1 million tons of oil equivalent in 1965 to 11,164.3 million tons in 2009[①]—in just 44 years, during which time the human population has more than doubled, our consumption of primary class energy has nearly trebled. It seems that this wanton energy consumption has brought about a series of environmental problems that may directly threaten the survival of mankind itself. Just one possible consequence of endless, unchecked energy consumption is accelerated climate change, which has recently been claimed to be caused by global warming.

As we experience more and more extreme environmental events, all human societies are coming to realize that more attention should be paid to protecting this planet's environment, not just to developments for our species. As a result, the United Nations Framework Convention on Climate Change (UNFCCC) was formed in 1992 as the first worldwide formal framework to settle problems brought on by climate change; but two decades later, the goal for reducing global greenhouse gas (GHG) still has a long way to go. The international community (with some exceptions, including China) reached a consensus in 1997 with the Kyoto Protocol. The target of reducing

---

① Data from *BP 2010 Statistical Review of World Energy*.

energy use based on emission control conflicted with fossil fuel energy-based economic development, thus, the original national emission reduction commitments became mere words on paper. In the past decade, however, the increased incidence of extreme weather has awoken world societies and their governments to the realization that the time left for human beings to solve this critical issue may well be running out. Therefore, at the 15th Conference of the Parties (COP15) to the UNFCCC in Copenhagen, 2009, after a seesaw negotiation, major emitters of greenhouse gases such as USA, China, the European Union and Japan finally made some concessions and announced emission reduction targets for the near future in their presentation at the conference.

With the pressure exerted by emission reduction targets, besides adopting the three market mechanisms Clean Development Mechanism (CDM), Emission Trading and Joint Implementation (all originally defined in the Kyoto Protocol) to meet emission reduction requirements, each major emitter also selected their policy options. These included developing clean energy, increasing investment in new technology, and/or changing the national industrial structure to confront the problem of conflict between emission reduction and social development.

Major emission countries also chose to introduce taxation policies, including a pigovian tax, which economists believe to be as effective for environmental control as environmental pollution control measures in helping to achieve the goal of reducing GHG emissions. Pigovian taxis most usually levied in the form of a tax on energy resource usage, or as a direct carbon tax related to amounts of carbon dioxide emission. Pigovian taxation has been used for many years in some industrialized nations as a means of environmental regulation. Denmark introduced a carbon tax in 1991, the world's first nation to do so; later, Finland, the Netherlands, Norway and Sweden also legislated for a carbon tax. Australia also adopted its Carbon Tax Act in October 2011, to become a further carbon-tax levying country among the industrialized nations.

China, a nation with the second-largest economy in the world, is also known to be one of the world's largest carbon emitters. China has continually been criticized for the weakness of its environmental policies and its irresponsible approach to environmental protection. Recently, a new report from the International Energy Agency (IEA) pointed out that "*China has now overtaken the United States to become the world's largest energy user."* It also said: *"China's demand today would be even higher still if the government had not made such progress in reducing the energy intensity (the energy input per dollar of output) of its economy."*

In fact, besides the universal condemnation by international society, 30 years of rapid economic development has caused China to be faced with serious damage to its own environment due to its energy resource consumption. *World Energy Report 2010*, in accordance with the above data, showed[①] that in 2009, China's coal consumption accounted for more than 49%—virtually one-half—of the world's total coal consumption, and 10.6% of total oil consumption, that year. The huge damage caused by such energy consumption is shocking: in Shanxi[②], China's leading coal region, nearly two million square kilometers of land has been excavated for open pit coal mining. This is one-eighth of the total area of Shanxi Province. The use of so much energy and resources has produced carbon dioxide, sulfur dioxide and other harmful substances on a largescale, which have directly brought about severely impaired air and water quality, and other issues. For water resources, the Chinese State of the Environment data for 2009 shows that in China's three largest freshwater lakes, one was Grade V (on a scale of I to V, Grade V being lowest quality) and the other two were worse than Grade V. For air quality, the same source showed that sulfur dioxide emissions in China in 2009 reached 22,144,000 tons, with 8,472,000 tons of soot emissions and 5,236,000 tons of industrial dust. These figures all indicate that today's China of high-speed economic growth faces many environmental problems as a result.

In the face of so much foreign and local criticism about the extent of environmental damage, the Chinese authorities have taken a series of steps to improve the situation: for example, many small, highly polluting and obsolete production operations have been shut down. In 2009[③], China cut off output from 60.06 million small thermal power units, equivalent to saving 6,400 tons of raw coal. In 2010, 88% of the 1,539 small collieries in China were closed in an attempt to limit potential pollution damage.

China remains opposed to the Kyoto Protocol, but sustainable development

---

① Details can be found in the *World Energy Report 2010*.

② *Beijing News*, December 14, 2010.

③ Report in *Economic Observer Net*, October 25, 2010.

needs, as well as the extreme pressure brought to bear by other nations, to reduce pollutant emissions, resulted in China's 2010 announcement of a pigovian tax in an attempt to achieve its COP15 commitment (that is, a 45% reduction in carbon dioxide emissions per GDP unit by 2020 compared to the year 2005). For this purpose, the pigovian tax program is as follows: in 2011, a 5% *ad valorem* energy resource tax was trialed in China's western provinces for one year, after which it was extended nationwide. Furthermore, China has announced its intention to introduce a carbon tax in 13 provinces and cities in 2013 and nationwide in 2015. In this context, it is now of practical significance to be able to predict the environmental protection benefits and economic losses resulting from a carbon tax.

In this study, a multi-regional computable general equilibrium (MRCGE) model linked to the Global Trade Analysis Project (GTAP) was constructed to predict the effectiveness of imposing a carbon tax in China and worldwide by examining its ramifications, both regionally and globally. This paper is divided into five further parts. Part 2 is a review of the literature on multi-regional CGE models and CGE studies on climate change issues both within and outside China. Part 3 is a summary of international response to the effects of climate change, and the action that China intends to take. In part 4, a static MRCGE model is described for examining the economic and environmental effects of energy resource taxes on different regions in China. The model is expanded in part 5 to link with the GTAP to evaluate the effect of carbon tax in regional China and in different regions worldwide. Finally, part 6 presents concluding remarks for the overall study.

Literature Review

2.1 What is CGE Model?

As stated earlier, this study mainly applied the research methodology of computable general equilibrium (CGE)[①] model analysis, used to analyze the likely economic and ecological effects of imposing an energy resource tax or a carbon tax. CGE models, based on neoclassical microeconomic theory, are widely used in macroeconomic analysis. A series of algorithms describing the behavior of various economic entities are used to examine the possible influence (s) of external policy actions to be imposed upon an existing equilibrium economy.

CGE models usually consist of two parts: one part comprises the algorithms describing the model system, and the other is a detailed database which is consistent with the algorithms; such databases are frequently prepared from input-output tables or social accounting matrices.

CGE models originated from the input-output technique first developed by Wassily Leontief, in which cost played a dominant role. The focus of modern CGE models used on a nationwide basis usually depends, more so than the Leontief model, on the level of economic development. For least-developed and developing countries, for example, the CGE model by necessity is generally constructed either by foreign researchers from a developed area or by native-born researchers who have studied abroad, and such models tend to focus on the constraints of shortages in skilled labor,

---

[①] A detailed description of the CGE model appears in *Wikipedia*: <http://en.wikipedia.org/wiki/ Computable_general_equilibrium>.

capital, foreign exchange and so on.

CGE modeling of developed economies originated from Leif Johansen's 1960 multi-sectoral growth (MSG) model of the economy of Norway, and also the Cambridge Growth Project (1960–1987, University of Cambridge) in England, which developed a model of the British economy. These models are pragmatic and dynamic (variables are traced through time). The Monash University (Australia) model is typical of this class. Modern CGE models may perhaps be traced back to that of Taylor and Black (1974).

CGE models are very helpful for estimating the impact of changes in one sector upon other sectors. For example, a tax on flour may affect the price of bread, the CPI, and hence perhaps wages and employment. As well as being widely used to analyze international trade policy for a country or a region, CGE models are increasingly being used to estimate the economic effects of reducing greenhouse gas emissions, or dealing with water shortage, and similar issues.

There are usually more variables than equations in CGE models; the excess variables must be therefore determined outside the model (exogenous variables). The other variables are determined by the model itself (endogenous variables). Different outcomes of the model are usually the result of different choices of exogenous variables, which may give rise to controversial research results. For example, some modelers maintain employment and trade balance as fixed values, while others allow these to vary. Exogenous variables usually include those that define technology, government instruments (e.g. tax rate) and consumer taste.

There are now many CGE models for different countries. One of the best known is the GTAP model (from the Global Trade Center at Purdue University, USA) for research into world trade economics.

CGE models may be either static or dynamic. Static models simulate conditions at a single point in time, and are generally used to simulate the result of an exogenous shock at a given time compared to the baseline conditions, or when other after-shock results require analysis. Thus, in static CGE models the process of adjustment to the new equilibrium is not given, only the outcomes (e.g., whether labor supply or share prices will adjust or not). This allows model builders to distinguish between short-term and long-term equilibriums.

Dynamic CGE model strack the changes in each variable through time, typically at annual intervals. These models are much more accepted, since they simulate economic activity more realistically. There are more limitations to building such a model, however, and the resulting simulation is more difficult to analyze for the reason that changes in the model's own parameters may be caused not only by exogenous (policy-driven) shock, but also by the model's own dynamic elements.

Recursive dynamic CGE models solve the problem for a single given time period, based on the idea that the behavior of the economy at one period depends only on its current and past states.

2.2 CGE Analysis on Carbon Tax Effect Worldwide

The CGE model methodology has been widely used to evaluate the economic and environmental effects of climate policies all over the world. In one example, Bergman (1991) established a CGE model to evaluate the effect of different environmental policies in Sweden. This indicated that major emission reductions in Sweden might have general economic equilibrium effects, and that emission control cost functions that failed to take these effects into account might give a distorted picture of the economic impact of emission control.

In Ireland, Jensen et al. (2003) used a disaggregated CGE model based on the well-known Australian ORANI model built by Dixon et al. in 1980, to analyze the effects of agricultural policy change regarding GHG emissions from the Irish agricultural and forestry sectors. They found that decoupling leading to agricultural GHG emission reduction would mainly occur by reducing the attractiveness of beef farming, and also that intermediate decoupling of agricultural payments would slightly under-achieve the National Climate Change Strategy (NCCS) target of 10% GHG reductions in agriculture, whereas full decoupling of agricultural payments would slightly over-achieve this target. The simulations did not incorporate technology effects on emission reduction, which suggested that it would be possible to set a more ambitious target for emission reductions from the agricultural sector. In a further stage of development, the IMAGE land economy model (Trinity College, Dublin) was used to assess the efficiency of the overall target for emission reductions in the agricultural sector compared to non-agricultural sectors of the economy.

Againin Ireland, Wissema and Dellink (2007) developed an applied general equilibrium (AGE) model specifically to analyze the impact of a carbon energy tax on Irish economy, based on taxation and energy use. The main findings of their research were that a carbon tax of around 10~15 euro per tonne of CO_2 would achieve a 25.8% reduction of carbon dioxide emissions in Ireland compared to the 1998 level. Furthermore, they found that a carbon tax would lead to a greater emission reduction than an equivalent uniform energy tax.

Bernard et al. (2003) formulated GEMINI-E3, a multi-sectoral, multi-regional CGE model, to assess the economic impact of Swiss climate policy in line with the Kyoto Protocol. Their research concluded that a uniform carbon tax of 313 USD per tonne of carbon ($/tC) would allow Switzerland to reduce CO_2 emission by 15% compared to 1990 levels by 2010. The research also concluded that Switzerland would benefit from joining a EU-wide emission trading regime, where importing half of the required emission reduction at $146/tC would assist Switzerland to reach its Kyoto emissions target.

In Germany, Böhringer and Rutherford (1997) constructed a CGE model to analyze the welfare cost of exemptions in environmental policy together with the issue of unilateral carbon taxes in an open economy such as Germany. Using 1990 data for West Germany, they found that (a) welfare losses associated with exemptions would be substantial, even when the proportion of exempted sectors in the overall economic activity and carbon emission is small; and(b) political pressure from labor unions and industries was alike because many of these sectors would suffer a decline in employment numbers, output and exports as a result of a carbon tax. They furthermore believed that, since atmospheric carbon dioxide was a global problem, unilateral carbon reduction in one country did not induce significant changes of carbon emissions elsewhere.

Nations in continents other than Europe also use CGE modeling to analyze the climate policy. In Japan, Ban (2007) used an "overlapping generations/infinite live agent" (OLG/ILA) version of the CGE model to examine the effect of carbon tax on the Japanese economy. This showed that a carbon tax of 1780~2477 yen/tC would lead to a 6% reduction of carbon dioxide emission compared to the 1990 level. It would also create a 0.98% decrease in Japan's household utility.

Also in Japan, a multi-regional CGE model by Hayashiyama and Abe in 2011 evaluated the effect of greenhouse gas discharge reduction policy for 47 prefectures and 15 industries, showing that the efficiency of carbon dioxide reduction policy decreased as carbon tax increased. At the same time, energy-intensive industries are the most affected by such policy, introducing a problem of regional fairness from the viewpoint of household utility in different prefectures.

In the USA, Kamat et al. (1999) used a CGE model to determine the effect of a carbon tax on the economy of the Susquehanna River Basin area to investigate its regional effect. Results indicated that a carbon tax of 16.96 $US/tC would have a rather negligible negative impact on the region's economy as a whole and the impact on GRP would be very small (from 0.01% to 0.14% of base year values), but the negative impact on its energy industries could be sizeable.

In New Zealand, Frank Scrimgeour et al. used a CGE model in 2005 to address important questions particular to carbon, energy and petroleum taxes. The model was focused on the energy sector, and allowed for substitution between various sources of energy and between energy and capital. Their results showed that an energy tax based on the energy content of fossil fuel would have the effect of reducing carbon emissions, but would be less effective than a carbon tax; however, capital stocks could be adversely affected by policy instruments such as carbon tax, leading to reductions in GDP, household consumption, exports and investment.

Finally, in Chile, Dessus and O'Connor (2003) created a CGE model to examine certain previously neglected climate policy benefits to the health of the population of Santiago, Chile. They found that, using even the most conservative assumptions, a carbon tax could reduce its CO_2 emissions by almost 20% from the 2010 baseline with no net welfare loss, but with a 10% reduction being closer to "optimal". If, instead, Chile were to target a 20% reduction in particulate concentrations, a "particulate tax" would incur slightly lower costs than an equivalent carbon tax to achieve the same health benefits.

The above examples demonstrate that, over the past two decades, CGE modeling has been widely used to analyze the climate policies of many nations.

2.3 CGE Analysis on Carbon Tax Effect in China

In recent years, since researchers in China have gradually become familiar with CGE models, many have begun using the technique to study the effects of China's energy resource fiscal policy changes on the nation's economy, or similar issues. For example, The study of the relationship between carbon tax and carbon emission reduction of He et al. (2002) showed that to achieve reductions of 10.5%, 15.5%, 20.5%, 24.5% and 30.5%, China's marginal abatement costs would be, respectively, 88.4 RMB/tC, 146.6 RMB/tC, 219.9 RMB/tC, 289.4 RMB/tC and 418.2 RMB/tC. Wang et al. carried out a similar study in 2004 in which a one-country static CGE model was established to simulate the effect of an *ad valorem* tax on energy resources in China for the whole of 2010. This predicted that China would lose between 0 and about 3.9% of GDP for a carbon emission reduction rate between 0 and 40%, and that the predicted marginal abatement costs across the nation would be about 100 RMB/tC for a reduction rate of 10%, and 470 RMB/tC for a reduction rate of 30%.

Based on the model of Xie and Saltzman (2000), in 2009 Wei Weixian established a CGE model with environmental feedback to predict the effect on China's economy of introducing an *ad valorem* energy resource tax. Five tax levels between 10% and 50% were simulated. Their results indicated that, besides the carbon emission reduction effect, an *ad valorem* tax on fossil fuels would lead to a reduction in several of the nation's microeconomic indices, such as GDP, household utility, employment rate, and so on, but that only an *ad valorem* tax rate more than 20% would cause all the microeconomic indices to decrease more than 1%.

Similarly, in 2009 Yang Lan et al. established a model that showed that an energy tax would have only a slight impact on China's economy (less than about 1%); at the same time, it would effectively change China's energy structure and reduce coal use in proportion to energy use. Besides these studies, Pang Jun et al., Xia Chuanwen and others have carried out related research on such issues as cutting fuel levy tax.

What all the above studies had in common is that they regarded the nation as the study object to illuminate the national response to exogenous economic shock. However, when a global problem occurs that affects a nation as vast as China, two features of the problem should be noted: (i) Since the country experiences extreme economic development differences from region to region, division on a regional basis should be considered when policy effects are being evaluated. (ii) Reduction in carbon dioxide emission is a global problem. No single country, however large, can ever successfully solve the problem on its own; cooperative action between countries and regions must be considered.

Some CGE models, however—for example GTAP (Hertel, 1997) and the Global Linkage model of the World Bank—incorporate data for multiple regions, as described by van der Mensbrugghe (2005). Such models divide the globe into areas linked by international trade. The multi-regional CGE model (MMRF-Green) developed at Monash University (Australia) considers the Australian states and territories separately. The model has been applied to a wide range of issues, including an analysis of greenhouse gas problems by Adams et al. (2000).

In the next section, two well-known MRCGE models, MMRF-Green and GTAP-E, are introduced.

2.4 Current Multi-Regional CGE Models

2.4.1 MMRF-Green Model[①]

MMRF-Green (Monash multi-regional forecasting) is a dynamic CGE model for analysis of polices related to the Australian economy. As a multi-regional CGE model, it allows for inter-fuel substitution in electricity generation by region, and the adoption of endogenous abatement measures in response to GHG control policy. The model structure of MMRF-Green model is described below.

(a) Structural overview

Australia's six states and two territories are modeled separately. The results from each model are then disaggregated to produce projections of output, employment and GHG emissions for 57 sub regions. Several kinds of economic agents are incorporated in the model, including domestic producers and investors for industries and regions, State and Territory governments, Common wealth (i.e., national) government, regional householders and an aggregate of foreign purchasers of exports and sellers of imports.

Each region has 40 industry sectors, each of which comprises one capital creator, a single household, a foreign purchaser, and a seller of goods and services. Government activity is captured at the regional level through State and Territory governments, and at the national level through the Common wealth government.

For the purposes of the model, all economic agents are regarded as operating in

---

① This part of model structure was referred to Jack Pezzey, Ross Lambie: CGE Model for Evaluating Domestic Greenhouse Policies in Australia: A Comparative Analysis, Consultancy Report.

perfectly competitive markets and engaging in optimizing behavior.

(b) Households

Household activities assume utility-maximizing activity in each model region. The structure of household demand is shown in Fig. 2.1.

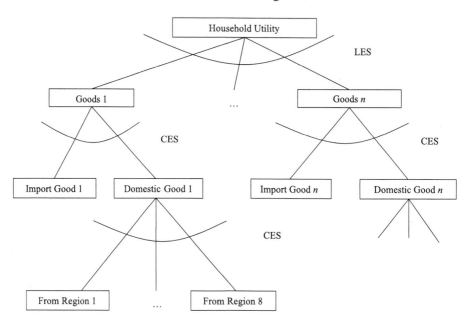

Fig. 2.1 Household Demand in MMRF-Green
Source: Pezzey and Lambie (2001), Figure 3.1, P.13

As the figure shows, the top level of household demand is regarded as a linear expenditure system (LES) for different kinds of consumer commodities, assuming that each household purchaser maximizes the utility but does not exceed their available budget. It should be noted, however, that the LES used in this model does not include luxury goods and services: instead, the Cobb-Douglas (CD) function is adopted to describe household budget expenditure on luxury commodities.

Four categories of regional household income are considered in the model: income from primary factors; employment-related income; income from government transfer; and income linked to the nominal GRP.

(c) Producers

Perfectly competitive markets and maximized profits using constant returns-to-scale technology are assumed for producers. Fig. 2.2 and Fig. 2.3 show how commodities are treated in the model.

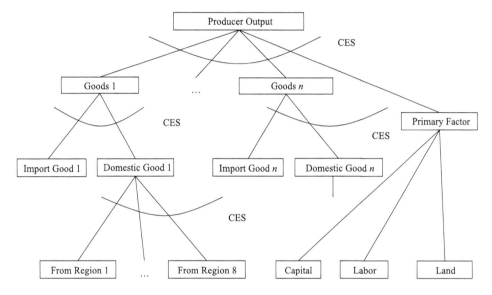

Fig. 2.2 Production Structure in MMRF-Green besides Electricity Sector

Source: Pezzey and Lambie (2001), Figure 3.2, P.14

For industries other than electricity supply, cost minimization is assumed for commodity production or service production for each producer. Production output is considered as a composite of intermediate commodities and primary factors, structured as a three-level nested production technology function. At the top level, all production from all industries are categorized as either energy-intensive or non-energy-related inputs, each category being produced from combined intermediate and primary input. For non-energy-related inputs, a Leontief production function is used to define the combination: that is, the highest substitution elasticity for these productions is fixed at zero. For goods emanating from energy-intensive input, the constant elasticity of substitution (CES) production function is adopted; that is, substitution elasticity is non-zero. Also, for energy-intensive goods, a distinction is made between the energy sources as being a petroleum product, electricity or natural gas, by setting higher substitution elasticity value as appropriate for the particular group of commodities.

The second level of the nested production technology function consists of a combination of imported goods and domestic supply using a combined CES function. Primary factor supply, as opposed to composite intermediate input, is a composite of capital, labor and land using a CES function. For the third or lowest level, domestic supply from the second level combines domestic production from eight regions using a CES function.

As shown in Fig. 2.3, the electricity supply industry is structured quite differently from the nested structure of the other production industries in Fig. 2.2. For each region, the model allows for electricity generation from any of five fuels (black coal, brown coal, gas, oil products and others) and, as with other products, the cost of each isused in a CES function to simulatea composite electricity supply. The model then assumes that the electricity supply sector takes all its supplies from all generator industries and sells the electricity to all other industries and endusers; thus the sole input to the electricity supply industry is considered to be the electricity generated from the different energy sources.

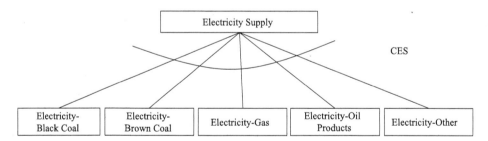

Fig. 2.3   Production Structure in MMRF-Green Electricity Sector

Source: Pezzey and Lambie (2001), Figure 3.3, P.15

(d) Capital creation and accumulation

The structure for the simulation of capital goods creation in the MMRF-Green model is similar to that for consumption goods creation in each region (Fig. 2.4); that is, it is assumed that capital producers minimize their production costs, that inputs comprise both domestic and imported commodities, and that a three-level nested structure is adopted. Fig. 2.4 also shows that no primary factors are used directly for capital goods production.

At the top level of the three-tier structure, commodity composites representing individual goods are combined in fixed proportions to produce industry-specific capital. At the second level, commodity composites are formed by CES function combination of domestic and imported goods; and at the third or lowest level, domestic goods from each region are combined by the CES function to simulate the composite domestic capital goods output from all domestic regions.

In the MMRF-Green model, capital demand depends on the analysis type. There are usually two types of static comparative analysis, over the short run (analysis period of 1~2 years) or the long run (analysis period >5 years). For short-run analysis, it is assumed that industry stocks of capital are fixed, and thus capital stocks in regional industries and national aggregate investment are exogenous. Differences in industry aggregate investments between regions are based on the relative rate of return on capital.

Unlike the short run, for the long run it is assumed that capital stock aggregation adjusts to maintain the rate of return for the whole economy. This rate of return is set exogenously. For both the long and short run, it is assumed that investment supply is perfectly elastic and that the capital market is open to international investors.

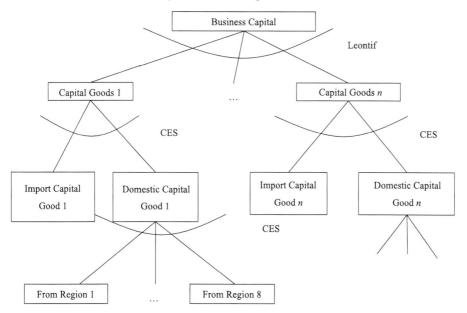

Fig. 2.4 Creation of Capital Goods in MMRF-Green

Source: Pezzey and Lambie (2001), Figure 3.4, P.16

(e) Government

Government activities in the MMRF-Green model include collection of taxes, receiving revenue from government-owned assets, making government transfers, and purchasing goods and services as a final user. For determining regional and central government demands in the model, there are three options available: (i) endogenous demand via specification of a rule; (ii) endogenous demand as an instrument that can yield an exogenously determined target; or (iii) exogenous demand. The main government income in this model is from taxes on income, sales and excise, import tariffs on international trade, and government-owned assets. Transfers can be from the government to the household sector or from central government to the regional government.

The central government policy and the macroeconomic environment are exogenous in the model. Moreover, government intervention in markets is as an *ad valorem* sales tax on target goods or services in the market.

(f) Trade and financial flows

The MMRF-Green model was described for regions in Australia and all international flows of goods and services are captured by export demand and import supply curves for each region. It is assumed that all domestic agents have a perfectly elastic supply curve for all imported commodities they trade. Export from each region in the model is divided into two different groups: traditional and nontraditional. Traditional exports include agricultural and mining products. All other exports are defined as nontraditional. Both types of commodity exhibit downward-sloping constant elasticity demand functions.

The specification described by Armington (1969) is adopted for international trade for every region in the model. That is, commodities and services for each region are substitutes for the same thing obtained from other regions or from international trade. It is assumed that central government adjusts policies to reach a trade balance. Under this assumption, investment and consumption in the model are proportional to changes in aggregate domestic expenditure. The elasticity used for both import and export trade is derived as the weighted average of the elasticity for the 115 commodities used in the Monash model.

(g) Labor market

For the whole nation, labor supply is determined by labor demands and demographic factors. At a regional level, the labor demand for each region is derived from choosing the lowest-cost combination of occupation-specific labor inputs from the producer side.

Regional labor movement is accepted in the model. Variables affecting the regional labor market are regional labor supply, unemployment rates and regional wage differentials; in the MMRF-Green model, any two of these variables can be set exogenously, thereby determining the third.

(h) Population growth

Three forces determine the population of a region in the MMRF-Green model: (i) natural population growth; (ii) foreign migration; and (iii) interregional relationships. The regional population is either exogenous with one endogenous regional labor market variable, or endogenous with exogenous setting of all regional labor market variables.

(i) Data

Data used in the MMRF-Green model comprise a multi-regional input–output (IO) table generated from disaggregation of the national IO table. This IO table was also used in the Monash model. The MMRF-Green model also shares primary factor values and elasticity of domestic-import substitution with the Monash model.

Regional data for government revenue and expenditure, demographics, employment and the labor force are from data published by the Australian Bureau of Statistics.

2.4.2 GTAP-E Model

(a) Model introduction

The GTAP-E model is based on the GTAP model constructed by the Global Trade Center of Purdue University (Lafayette, IN, USA). It differs from the original GTAP model by focusing on an economy-energy-environment (3E) linkage and uses current data on energy consumption and CO_2 emissions according to economic activities, with a top-down mechanism used to describe energy substitution. The main aim of Burniaux and Truong (2002) was to construct an effective model for analysis of energy control policies for different countries and of global GHG trading arrangements. Since the model focuses on 3E linkage, it is termed GTAP-E.

(b) Differences from GTAP

Domestic economic activities and international trade in the GTAP-E model do not differ much from the GTAP model. The only differences are several small improvements in the GTAP-E model, such as 3E linkage.

First, all energy consumption for economic activities is included in the GTAP-E model. There are five energy commodities: coal, oil, natural gas, and electricity and petroleum products. Consumption data for these resources were taken from the energy balance database of International Energy Agency (IEA).

Second, the GTAP-E model includes a carbon tax for each country (region). The carbon tax is levied on all economic agents (production sectors, private household sectors and government) and is used by the local government to fund their activities. As a policy tool for price control, it is not surprising that such policy increases the price of high-carbon energy resources such as coal.

Third, single energy substitution in the model is described by constant elasticity of substitution (CES) functions for easy identification of price changes caused by the carbon tax. In addition, the total energy consumption change for each economy can be observed.

Fourth, CO_2 emission is estimated using the Intergovernmental Panel on Climate Change (IPCC) methodology, as described by Houghton et al. (1997).

Finally, the GTAP-E model includes an international trading system for carbon emissions. Thus, the model can be used to evaluate the effects of different mechanisms and to identify suitable GHG reduction ranges for different countries or regions.

The GTAP-E model structure is shown in Fig. 2.5. The model details have been described by Burniaux and Truong (2002).

(c) Production sector

The GTAP-E model assumes that each industry sector has production activity combined with a primary factor input, an energy resource input and other intermediate factor inputs under a CES function. An agent of this sector seeks to minimize his costs under the zero-profit condition. The production function of this model comprises eight level nested structures, as shown in Fig. 2.5 and Fig. 2.6. One difference between the GTAP-E and GTAP models is a capital-energy composite input that is combined with the primary factors.

Literature Review 19

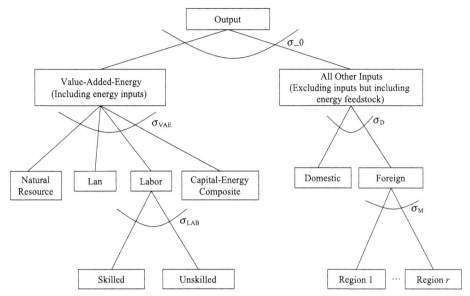

Fig. 2.5 GTAP-E Production Structure

Source: Burniaux and Truong (2002), Figure 16, P.31

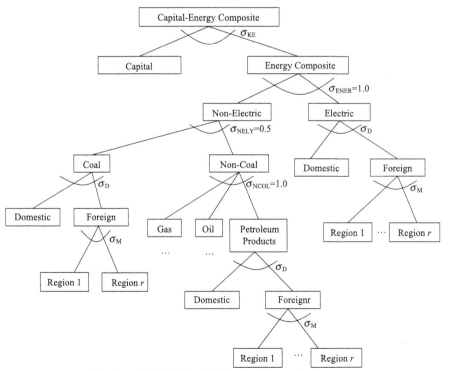

Fig. 2.6 GTAP-E Capital-Energy Composite Structure

Source: Burniaux and Truong (2002), Figure 17, P.31

The GTAP-E model involves two steps to describe energy substitution. First, energy resource factors are divided from the ordinary intermediate input; they are then combined to yield a composite energy input. The energy inputs divided were electric and non-electric, coal and non-coal, and gas, oil and petroleum products. The composite energy input has zero elasticity of substitution with other intermediate inputs. The supply for each energy resource follows the Armington (1969) assumption, because they are not completely substitutable commodities. The elasticity of substitution for domestic products with import products and exports products is σ_D and σ_M.

In the second step, the composite energy input is evolved to a capital composite for energy goods by including the capital input from the primary input sector. This combination is also subject to a CES function.

(d) Government

For the government and household activity sector, GTAP-E follows the same assumptions as the GTAP model, treated separately for the government sector and the household sector. Fig. 2.7 shows the consumption structure for government.

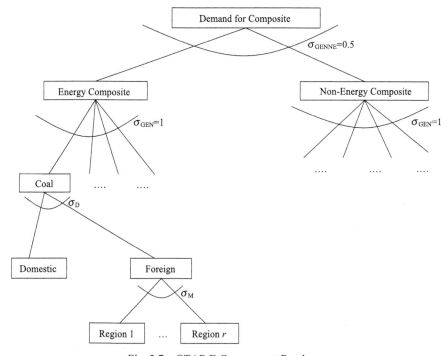

Fig. 2.7 GTAP-E Government Purchases
Source: Burniaux and Truong (2002), Figure 18, P.37

It is assumed that the government utility follows a Cobb-Douglas function. Under this function, the composite energy commodities are separated from non-energy commodities. In the energy composite, each energy commodity is substituted under a CES function.

(e) Household sector

Fig. 2.8 shows the private purchase structure for the household sector in the GTAP-E model. For consumption by private households, a non-homothetic constant difference of elasticity (CDE) function was used. For calculating changes in utility consumption by households, the per capita number was used as a baseline. Consumption of energy commodities follows a CES function, as for government consumption.

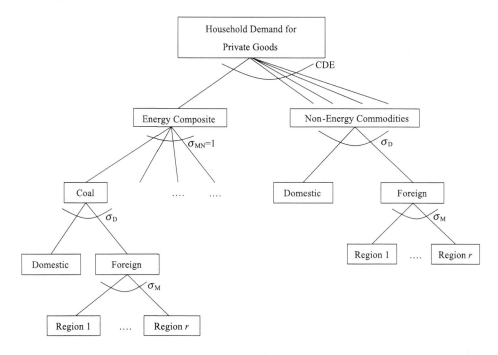

Fig. 2.8 GTAP-E Household Private Purchases
Source: Burniaux and Truong (2002), Figure 19, P.38

(f) Data

The newest GTAP-E model uses the GTAP 6.0 database, which is based on 2004 production and trade data. This database is substantially modified in areas such as the energy sector, and data on energy, emissions and population are added.

Emissions of CO_2, methane and nitrous oxide are represented as CO_2 equivalents, using global warming potentials of 1, 21 and 310, respectively. Data on CO_2 emissions from fossil fuel combustion were sourced from the International Energy Agency. Obtaining data on non-combustion emissions is problematic, so the relevant data are either estimated or compiled from the United Nations Framework Convention on Climate Change (UNFCCC) inventory data for individual countries.

CGE models such as MMRF-Green and GTAP-E are mature models that are widely used for policy analysis. However, these CGE models only consider regional divisions within one country or in world. In this study, we created a one-country multi-regional CGE (MRCGE) model linked to GTAP, which considers both regions within one country and global regions to evaluate the effectiveness of a carbon tax in China and worldwide.

However, before analysis of policies under the MRCGE-GTAP model, it is important to review actions carried out throughout the world and in China to prevent climate change from worsening.

Responses to Current Climate Change 3

3.1 International Responses

On June 4, 1992, countries around the world reached a consensus on the issue of climate change and signed the United Nations Framework Convention on Climate Change (UNFCCC). This convention, which formally came into force on 21 March 1994, was the first in history to deal with climate change. To date, 189 countries have signed up to the agreement. The first Conference of the Parties (COP) was convened in 1995 in Berlin, Germany. Since then, COP has been held 17 years running. The exact content of each conference is shown in Table 3.1.

Table 3.1 Details about Each COP Conference

Conference	Year	Place	Conference achievements
COP1	1995	Berlin, Germany	Meeting adopted the documents like "Berlin power of attorney"
COP2	1996	Geneva, Switzerland	Meeting did not get a consensus on the drafting of the protocol and decided to continue the discussion
COP3	1997	Kyoto, Japan	The Kyoto Protocol was adopted by representatives of 149 countries and regions at the meeting
COP4	1998	Buenos Aires, Argentina	Groups of developing countries differentiated into: the Alliance of Small Island, countries looking forward to the CDM, China and India
COP5	1999	Bonn, Germany	The Annex to the Convention was adopted, like guide for technical review of greenhouse gas inventories, etc
COP6	2000	Hague, Netherlands	The European Union, the United States and large developing countries (China and India) stood like the legs of a tripod in the negotiation
COP7	2001	Marrakech, Morocco	A package of high-level political decisions were made on the Kyoto Protocol compliance issues, Marrakech Agreement file was formed

Continued

Conference	Year	Place	Conference achievements
COP8	2002	New Delhi, India	The Delhi Declaration adopted at the meeting emphasized that reducing greenhouse gas emissions and sustainable development were still important tasks of the States Parties in the future performance
COP9	2003	Milan, Italy	In the case of U.S. withdrawal from the Kyoto Protocol, Russia refused to ratify the Protocol, resulted in the Protocol could not be effective
COP10	2004	Buenos Aires, Argentina	Representatives from over 150 countries discussed around the achievements of the United Nations Framework Convention on Climate (UNFCCC), and the challenges that the UNFCCC faced with in the future
COP11	2005	Montreal, Canada	February 16, 2005, the Kyoto Protocol entered into force. In November, more than forty important decisions were reached in the conference
COP12	2006	Nairobi, Kenya	Dozens of decisions were reached, including the Nairobi work program. Agreement was achieved on the problem of management of the Adaptation Fund
COP13	2007	Bali, Indonesia	The meeting focused on a commitment on how to further reduce greenhouse gas emissions after the expiration of the Kyoto Protocol in 2012. The Bali roadmap was adopted
COP14	2008	Poznan, Poland	Group of Eight leaders reached agreement on long-term greenhouse gas emission reduction targets, which were together with other States Parties to reduce at least a half of the global emissions by 2050
COP15	2009	Copenhagen, Denmark	Developed and developing countries had fierce conflict in meeting, and finally reached the political will which were not legally binding documents—Copenhagen Protocol
COP16	2010	Cancun, Mexico	Two resolutions, the Convention and the Protocol were adopted. The greatest achievement of the meeting was that the United Nations climate negotiation conference was saved
COP17	2011	Durban, South Africa	The conference was agreed to execute second period of Kyoto Protocol, but Canada had announced to quit the Kyoto Protocol through this meeting

Data source: Collected from the previous sessions of the COP content

Of all of these conferences, COP3, held in Kyoto, Japan, in 1997, took on landmark significance, since it obliged all Parties to agree on the Kyoto Protocol. The Kyoto Protocol formally came into force eight years later, in 2005, when it became the first legal document in history to impose legal restrictions on greenhouse gas emissions.

The objective of the initial stage (2008 to 2012) of the Kyoto Protocol was to reduce the major industrialized nations' emissions of six greenhouse gases, including carbon dioxide, by an average of 5.2%, calculated using the year 1990 as its baseline. Of these, the European Union was to reduce emissions by 8%, the USA by 7%, Japan by 6%, and Canada by 6% and other eastern European countries by 5%~8%. Meanwhile, countries whose greenhouse gas emissions were already low, such as New Zealand, Russia and Ukraine, were to keep emissions stable in relation to their 1990 level. Simultaneously, the Protocol permitted Ireland, Australia and Norway to increase their emissions by 10%, 8% and 1%, respectively, in relation to the 1990 levels. These restrictions were on the emissions of the relevant developed countries, although there were no specific legally binding provisions in the Protocol for developing countries, including China, to reduce emissions. However, the unhappy truth is that the objectives set out for stage one of the Kyoto Protocol have not remotely been achieved. In fact, up until the 2011 round in Durban, South Africa (COP17), the Parties charged with reducing emissions continued their gambling game. In this high-stakes international game, two principal points lie at the heart of the debate.

First, whether developing countries, including China, need to accept demands to reduce emissions. Developed countries believe that where international cooperation is concerned, China and India, two developing countries with appreciable absolute values, ought to accept demands to reduce emissions, and, moreover, the emissions process must be "quantifiable, accountable and verifiable". However, developing countries, including India and China, continue to call for equitable global development based on a principle of "collective but differentiated responsibility". The focus of this debate is that developing countries in large part believe that the high level of greenhouse gases emitted by developed countries in the course of 200 years of industrialization is a major factor in present-day climate change. Therefore, they argue that it ought to be the historical duty of these countries to take on the bulk of the responsibility for reducing emissions. Furthermore, in the eyes of developing countries, developed countries have already seized an excessive proportion of the world's limited atmospheric resources, and thus diminished the space available for emissions, which many developing countries rely on as a basic living requirement. Moreover, many developing countries remain in the initial stage of industrialization, so they argue that the mooted collective reduction of emissions essentially limits their development potential. In their eyes, such restrictions are neither fair nor rational.

From a statistical point of view, there is a certain amount of sense in the claims of the developing countries. According to the statistics available from the World

Resources Institute (WRI) in the USA, over 70% of existing man-made greenhouse gas emissions come from developed countries. Statistics also suggest that between 1850 and 2005, a total of 1,122.2 billion tonnes of carbon dioxide were emitted worldwide. Of this amount, emissions from developed, industrialized countries accounted for 806.5 billion tonnes, which equates to over seven-tenths of overall emissions worldwide.

During this period, emissions from the European Union alone accounted for 27.5% of the total amount. Per capita, accumulated emissions came to 542 tonnes in the European Union, 958 tonnes in Germany and 1125 tonnes in the UK. Worldwide, emissions equalled 173 tonnes per capita. From a statistical standpoint, the complaints of developing countries are indeed based in logic. However, it is important to note that, in comparison with past emissions, present-day absolute emissions levels remain a problem that must not be overlooked. Take the absolute emissions levels of the major international nations in 2009 as an example. As for that year, the absolute emissions levels of the developing countries India and China are already the highest in the world. Therefore, it can be seen that if responsibility for reducing emissions must fall only to those with the highest greenhouse gas emissions in the past, the objective of reducing greenhouse gas emissions worldwide will continue to be difficult to achieve.

Leaving aside the question of who should take responsibility for the further reduction of emissions, the second topic of debate over global reduction of emissions is that developing countries widely doubt the credibility of developed countries' undertakings to developing countries in terms of financial support and technology transfers. At every round of COP, there are calls from developing countries for developed countries to provide these things to enable the former to build suitable mechanisms and systemic protection. Indeed, technological innovation and transfer is critical to achieving development and countering climate change. Industrialized nations have accumulated a wealth of climate-friendly technology and management experience during the process of industrialization; only if developing countries are able to make full use of these existing climate-friendly technologies, and then research and develop new, low-carbon technologies, will they be able to implement the greatest possible reduction in greenhouse gas emissions on an international scale at relatively low cost. They may even be able to open up the green economy, low carbon industry as a new sector—a new ray of light—in the collective international response to climate change.

In fact, as far as the above mentioned second point of debate is concerned, the argument of the developing countries does not fully hold water, realistically. Since the Kyoto Protocol came into force, nations have been able—according to the stipulations of the Protocol, and in order to promote its objective of reducing every country's greenhouse gas emissions—to utilize green development mechanisms in international interaction, thus promoting cooperation between developed and developing countries to reduce greenhouse gas emissions. Under the "Restricted Allowances and Trade" system, buyers can purchase emission-reduction quotas which are determined and allocated (or auctioned) by the administrator; customers purchase emission-reduction quotas from projects that can reduce greenhouse gas emissions. This is the Clean Development Mechanism (CDM). The majority of developed countries use CDM trading in various guises to fund investment support and technology transfers to developing countries.

The advent of the Clean Development Mechanism opened up a vast carbon trading market between developed and developing countries. According to data published on the website of the UNFCCC, by May 2011 there were over 6000 CDM projects worldwide successfully registered by the executive board of the UNFCCC; furthermore, over 2600 projects were pending audit and authorisation. These registered projects were equivalent to authorising a reduction of 600 million tonnes of carbon dioxide emissions. Data on CDM from 2008 showed that the majority of buyers active in CDM trading markets were from Europe, Japan and Canada. According to statistics, of the different categories of buyer operating in CDM trading markets, 34% were funds, 58% private enterprises and 8% governments. The most active buyers in the market came from the UK, and the majority of these were part of private financial institutions; next were Italian electricity companies. This demonstrates that in recent years, the various financial and technological support provided by developed countries to developing countries has not, in fact, dried up; it just may not have been as generous as the latter might have wished.

In truth, many large multinationals in industrialised nations have, in addition to actions carried out in accordance with existing global agreements, and made a certain effort within their spheres of influence to tackle the issue of reducing greenhouse gas emissions.

3.2 Nation and Regional Political Responses

On a national and regional political level, developed countries chiefly use legal means to enforce restrictions on the emission of greenhouse gases, and methods such as fiscal policy to help realise reductions in emissions. One example is the USA, which has already announced its intention to withdraw from the Kyoto Protocol, and whose Senate released the draft "U.S. Energy Act" in May 2010. This act stipulated that, taking 2005 emissions levels as its baseline, US greenhouse gas emissions should be reduced by 17% by 2020, by 42% by 2030, and by 83% by 2050. Additionally, at least 140 US cities assert that they remain on course to fulfil the Kyoto Protocol's target of a reduction in emissions of 5% from 1990 to 2012. Among them is California, which announced that it will reduce emissions by 25% by 2020, and also included in the bill a stipulation that every car in the state must reduce greenhouse gas emissions by 30% by 2016.

The European Union, meanwhile, has already implemented its second European Climate Change Program (ECCP II), which represents a significant step in realizing reductions in greenhouse gas emissions according to the Kyoto Protocol. The European Union has undertaken to reduce greenhouse gas emissions by 8% by 2012, in relation to their 1990 level. The European Parliament has called on member states to improve energy efficiency by at least 11% in the next decade.

In August 2010, Germany, not only the pre-eminent nation in the European Union but also the member with the highest greenhouse gas emissions, passed a bill entitled "National Plan of Action for Renewable Energy". This plan stipulated that by 2020, 18%~20% of energy consumption in Germany should come from renewable energies. In September 2010, the German government issued another bill, entitled "Energy Planning: an Environmentally Friendly, Reliable and Affordable Energy Supply", which is a general strategy for renewable energy development until 2050. Germany's greenhouse gas emissions are due to fall, in relation to 1990 levels, by 40% by 2020, 55% by 2030 and 80%~95% by 2050.

In 2010, another member of the European Union, France, released a communique via its environment and sustainable development ministry which included details of the French government's "Natural Cities" blueprint, due to be implemented with immediate effect. It set out a number of measures to reduce pollution and improve urban environments, including reducing energy consumption, more rigorously managing water, air, and noise and litter pollution, restricting the use of chemicals, and expanding natural urban scenery in terms of both quality and quantity.

Legislation aside, the types of fiscal and administrative policy adopted by every developed country include: natural resource agreements, taxation on energy, taxation on carbon dioxide, emissions trading, production quotas for renewable energies or combined heat and power (CHP), and energy efficiency standards, as well as direct financial incentives to encourage the use of renewable energy, such as preferential rates, donations and tax exemption. These policies have been constantly updated according to circumstances. Taking the example of taxation on resources and carbon dioxide, initially the method involved taxation alone, but recently it was changed to a "taxation plus subsidies" method.

Since the 1990s, some developed countries have introduced taxation on energy, or on carbon dioxide with a base of carbonaceous fuel, in order to increase tax revenue and reduce their reliance on foreign oil supplies. Owing to the effectiveness of this form of taxation in reducing energy consumption and greenhouse gas emissions, many developed countries adopted it as an important means of decreasing greenhouse gas emissions. Yet within a short time, some countries offered low taxation rates to large energy-consumers in an attempt to prevent carbon dioxide taxation hampering the competitiveness of their national industry in international markets. For example, Norway, which introduced this form of taxation at a relatively early stage, reduced the rate of taxation on carbon dioxide produced by offshore oil and gas; in the Swedish manufacturing industry, the rate of taxation on carbon dioxide has been aligned to the standard tax rate of 35%, while in some energy intensive industries the rate of taxation has been reduced to almost zero; in the UK, the taxation rate on carbon dioxide currently stands at the original standard rate of just 20% in energy intensive industries. In order to promote technological development and avoid harming the competitiveness of national industries in international markets, numerous countries replaced taxation with subsidization, and implemented tax reduction policies or tax breaks on highly energy efficient methods such as renewable energies and CHP in order to boost supply and consumption. From the supply side, this was mainly involved fiscal measures related to renewable energy production or CHP, including the reduction or abolition of production tax, fixed assets tax, VAT, import duties and others.

For instance, the UK government adopted a fiscal policy designed to aid the development of CHP. In 2002, the UK's CHP output was 4700 MW; to meet the government's objective, the UK needed to establish 10,000 MW worth of highly energy efficient CHP output by 2010. To this end, the UK government exempted the CHP industry from climate change tax, and created fiscal incentives in the form of investment subsidies to enterprises investing in CHP.

In France, corporate taxation of CHP enterprises was lowered by 50%, while some municipal governments were allowed to reduce this rate as much as 100%. Outside the CHP industry, France introduced fiscal policy favoring the use of renewable energies. By virtue of tax breaks and a reduced VAT rate, companies saw their renewable energy equipment costs fall by 15%; another policy offered companies investing in renewable energies a significant reduction.

3.3 Responses from Multinational Cooperations

Leaving aside national and regional political intervention, since it is considered that the need to decrease greenhouse gas emissions largely stems from energy consumption, many natural-energy or energy intensive multinationals participate in efforts to reduce emissions in accordance with the United Nations Framework Convention on Climate Change.

For example, GE, a relatively energy intensive company, signed up to the UNFCCC in July 2005 with the self-imposed aim of reducing greenhouse gas emissions by 1% by 2012. It also doubled its investment in research and development to USD 1.5 million, in order to become more energy efficient by way of technological improvement.

DuPont reduced greenhouse gas emissions by 40% between 1994 and 2000. Earlier it had pledged to reduce greenhouse gas emissions by 65% by 2010, in relation to the 1990 level, as well as setting other objectives including keeping its total energy usage in 2010 at the 1990 level, and sourcing 10% of its total energy from renewable by 2010.

The world's biggest fine chemicals manufacturer, Rohm and Haas, pledged that by 2010 it would reduce energy consumption by at least 1%, as compared with the 1994 level, for every pound of products it manufactured. Novartis pledged to improve its energy efficiency by 2% annually.

Between 1996 and 2005, Dow reduced by over 20% the amount of fossil fuels it used to manufacture every pound of chemical products. Over the decade from 2006 to 2015, Dow set an objective of improving its energy efficiency by 25% and reducing greenhouse gas emissions by 2.5% annually.

BASF has taken a range of measures to reduce CO_2 emissions. In 2004, it used steam from the waste heat generated in the production process, which accounted for 48% of its total steam requirements. The company installed two combined cycle gas turbines at its plant in Ludwigshafen, Germany, which can produce over 3.5 times more electricity per tonne of steam than conventional combined systems, yet CO_2 emissions fell by over 500,000 tonnes per year. In another project, it used a more advanced catalyst system which significantly improved the production rate of acrylic acid, such that BASF was able to reduce CO_2 emissions by 230,000 tonnes per year. These and other projects enabled the company to reduce greenhouse gas emissions by 38% by 2002 in relation to the 1990 level. BASF is currently implementing its objective of reducing emissions by a further 10% by 2012 in relation to the 2002 level.

At the 2005 Montreal Climate Change Conference, Bayer received the "Low Carbon Leaders Award" for its accomplishments in the field of climate protection, and it was also crowned as "Leader in its Class" by experts appointed by the International Climate Change Initiative. It received the awards because in the decade between 1997 and 2006, as the company achieved a marked increase in production, its greenhouse gas emission levels nevertheless fell sharply. Bayer reduced its greenhouse gas emissions by streamlining its production process and deploying advanced technology, as well as closing several factories and selling some affiliate and subsidiary companies. Besides, Bayer effectively used gas power plants instead of coal plants, ensuring energy supplies through a combination of heat and electricity. This series of techniques enabled Bayer to achieve a reduction of over 60% in greenhouse gas emissions, a process it began in the early 1990s.

Energy intensive enterprises aside, a number of energy companies are currently actively taking measures to reduce CO_2 emissions. For instance, BP has built a plant in Peterhead, UK, at which the CO_2 in natural gas is pumped back underground and used to improve the oil recycling rate. According to information released by BP, this measure, along with various other measures and various energy efficiency projects, are destined to enable the company to meet its target of reducing greenhouse gas emissions by 10% in relation to the 1990 level.

Another famous energy company, Shell, set an objective of reducing greenhouse gas emissions by 5% by 2010 in relation to the 1990 level. In order to realize this objective, Shell subdivided the objectives of each of its activities into quantifiable sections; for example, its Canadian subsidiary pledged to reduce greenhouse gas emissions by 6% by 2008 in relation to the 1990 levels in its oil refining, exploration and production divisions. Greenhouse gas emissions in its petroleum sand division were due to fall 50% by 2010.

Apart from the pledges and actions of individual companies, some cooperative business agreements relating to the reduction of greenhouse gas emissions are also being concluded. For example, in 2007, 10 US companies collaborated with four environmental groups to promote the establishment of an agreement to reduce greenhouse gas emissions, in which they pledged to restrict emissions to 100%~105% of the level at that time after five years, to 90%~100% of the 2007 level after 10 years, and to 70%~90% of the current level after 15 years. By 2050, they aim to decrease emissions to 60%~80% of the 2007 level. Simultaneously, these companies announced that they would promote the use of energy efficient technology, and restrict or control CO_2 emissions. Coal-fired electric power plants would be the focus in terms of reducing emissions, and CO_2 would be captured before being pumped back underground. The parties to this agreement included Alcoa, BP, Duka, DuPont, FPL Group, GE, Lehman Brothers, PG&E and PNM, as well as an environmental safety body, a natural resources safety committee, the Pew Center on Global Climate Change and the global Natural Resources Institute.

From this variety of existing reports and materials, it is evident that developed countries and their powerful multinational energy suppliers or energy intensive companies have made pledges and taken steps of all kinds over the last 20 years. However, the thing worth noting among all of the various information that has been traced is that various news items give predominance to the pledges made, whereas there are no more materials offering insight into which of them were actually implemented. As a result, as to the actual results of the many measures adopted by developed countries to reduce emissions, perhaps more time is necessary in order to provide evidence.

3.4 China's Responses

China is the world's largest developing country. It has a huge population but a fragile ecology, and there is a great regional imbalance in development. In 2010, China's GDP surpassed CNY 29,000 per capita. According to the United Nations' definition, 100 million Chinese still live in poverty. Developing the economy, eradicating poverty and improving people's lives are vast, difficult tasks. However, despite China's attempts to move closer to solving these problems, the energy consumption, greenhouse gas emissions and other by-products of its 30 or so years of breakneck development have attracted international pressure from many quarters.

Faced with the situation described above, the Chinese government has, by and large, taken a positive stance in response to the external pressure and internal damage of various kinds caused by climate change. In practice, meanwhile, China has incorporated this major topic—one which affects economic and social development across the board—into its long-term economic and social development plans. As one of the countries not compelled by the Kyoto Protocol to take action, China unveiled, in 2006, a restrictive target of reducing GDP per unit of energy consumed by around 20% in relation to its 2005 level. In 2007, it became the first developing country to set out, and implement, a proposal for dealing with climate change, and in 2009, it confirmed its objective of reducing its GDP per unit of greenhouse gas emissions by 40%~45% by 2020 in relation to the 2005 level.

China's main development planning is broken down into five-year periods, for which general plans are set out. To date, 11 of these "Five-Year Plans" have been executed (China's 11th Five-Year Plan covered the period 2006 to 2010). In 2011, a new planning phase commenced: the 12th Five-Year Plan.

During the period of the 11th Five-Year Plan, China addressed the objectives it had set for itself. It used five means of controlling greenhouse gas emissions, guiding the reorganization of industry composition and energy composition, advising on energy-saving and improving energy efficiency, developing brand new low carbon energies and increasing the number of carbon sinks. It achieved a measure of success in this process.

In terms of reorganizing industry composition, China announced plans to reorganize and revive 10 key industries, including the automotive and rail industries. It revised its "Guidelines for the Reorganization of Industry Composition" and released "Ideas for Suppressing Excess Production and Replica Construction in Some Industries to Induce Healthy Development". It raised barriers to entry in high energy-consuming industries, examined and evaluated energy-saving measures in fixed-asset investment projects, improved the redesigning and upgrading of technology in traditional industries, promoted company acquisitions and restructuring and revised its policy on tax rebates on exports. It introduced export duties on coal, some non-ferrous metals, steel billet, chemical fertilizers and other products, and suppressed the exportation of natural resources and products consuming large amounts of energy or causing high emissions. Simultaneously, it adopted a strategy of expanding large-scale projects and closing small-scale projects, altogether shutting down 76.82 million kW of grouped low-fire electric generators, dispensing with outdated production techniques with the capacity to make 72 million tonnes of smelted steel, 120 million tonnes of smelted iron, 370 million tonnes of cement, 107 million tonnes of coke, 11.30 million tonnes of paper and 45 million weight boxes of glass. In the Chinese electrical industry, the percentage of grouped flame-powered electric generators as a proportion of total electrical installation capacity rose from 47% in 2005 to 71% in 2010; the percentage of iron smelting production capacity from large-scale (over 1000 cu m) blast furnaces in the iron and steel industry rose from 48% to 61%; and the proportion of large-scale production of pre-heating cells in the aluminum electrolysis industry rose from 80% to over 90%. Between 2005 and 2010, coal consumption in the supply of electricity from thermal power sources fell from 370 g/kW • h to 333 g/kW • h, or 10%; the energy used in steel manufacture fell from 694 kg per standard coal unit to 605 kg per standard coal unit, or 12.8%; the energy used in cement manufacture fell by 24.6%; energy used in ethylene manufacture fell 11.6%; and the energy used in synthetic ammonia manufacture fell 14.3%.

In implementing energy-saving measures, China has adopted more stringent procedures to investigate those responsible for meeting objectives. It has established a system for monitoring and investigation, and performs regular investigations and appraisals of 31 provincial governments and 1000 key enterprises with regard to completed energy-saving objectives and implemented energy-saving objectives. In 2010, China carried out strict investigations into accountability and responsibility for meeting objectives by specially supervising the extent of development of energy saving and emissions reductions in 18 key regions, thus promoting the achievement of energy-saving objectives nationwide.

Furthermore, China actively promoted the implementation of energy-saving activities in its key energy-consuming fields. Throughout the period of the "11th Five-Year Plan", China implemented 10 major energy-saving projects, including redesigning industrial boilers (furnaces), CHP, energy saving in electrical machinery systems and the use of residual heat and pressure. It rolled out energy-saving procedures in 1000 enterprises, stepped up energy-saving management procedures in key energy-consuming enterprises, and promoted the auditing of energy use and assessment of energy efficiency in relation to its objectives. Simultaneously, China increased the implementation rate of obligatory energy-saving standards in new buildings, accelerated the redesign of existing buildings in order to save energy, promoted the use of renewable energy in buildings, and redesigned the workplace residences of government officials in order to save energy. According to statistics from late 2010, in Chinese urban areas, the implementation rate of obligatory energy-saving standards in new-builds at the design stage was 99.5%, while at the construction stage it was 95.4%. In the period 2006~2010, the total area of energy-efficient buildings built in China was 4.857 billion sq m, and they had an energy-saving potential of 46 million tonnes of standard coal units.

Aside from stipulating investigation targets and enforcing restrictions on energy use in key industries, China has worked hard to promote energy saving; for example, by promoting low-energy technology and products, developing a circular economy, promoting energy-saving market mechanisms, and adopting policies that offer incentives to implement energy efficient procedures within China. With a view to promoting energy efficient technology and products, China published a three-part list promoting 115 key national energy efficient technologies and focusing on seven energy-efficient technologies in sectors including the steel and iron industry, the construction materials industry and the chemical processing industry.

Additionally, China has used financial subsidies to promote energy efficient products such as highly energy efficient lighting products and air-conditioning systems, and energy efficient electrical appliances. Through financial subsidies from central government, it has promoted 360 million highly energy efficient lighting products, 30 million highly energy efficient air-conditioning systems and 1 million low-energy cars, thus achieving an energy-saving potential of 20 billion kW • h per year. In terms of low-energy and new-energy cars, meanwhile, China has taken the lead in the public service sector by advocating the use of hybrid vehicles, electric cars and fuel cell cars. The period of the 11th Five-Year Plan saw the widespread application of a large range of highly energy efficient technologies, including pure low-temperature residual-heat energy generators, new cathode technology aluminum-lectrolysis cells, high-pressure frequency converters, rare-earth permanent-magnet electrical machinery, and plasma oil-free ignition systems. Such technology now occupies 67% of the lighting products market and 70% of the low-energy air-conditioning market.

As far as the circular economy is concerned, China rolled out a nationwide basic model for construction entitled "Urban Minerals". This extolled the large-scale, circular and high-value uses of discarded resources such as scrap electrical equipment, outdated household electrical appliances, scrap plastics and scrap rubber. By actively promoting industrial multipurpose utilization of solid waste products, within five years China reused some 1 billion tonnes of fly ash, 1.1 billion tonnes of waste rock and 500 million tonnes of metal smelting slag.

In terms of advocating low-energy mechanisms, China has promoted energy saving through market mechanisms such as energy management usage contracts, management of demand for electricity and voluntary agreements on energy saving. In 2010, it released an ordinance entitled "Ideas for Accelerating the Implementation of Energy Management Agreements to Promote the Development of the Energy Saving Services Industry". This document provided extra backing to the energy-saving services industry insofar as it increased financial support, introduced advantageous fiscal policies, improved the relevant accounting systems and polished financial service techniques. From 2005 to 2010, the number of energy-saving service companies in China jumped from 80 to 800, directly creating 165,000 new jobs; the energy-saving services industry increased in value from CNY 4.7 billion to CNY 84 billion, and consequently the annual energy-saving potential increased from over 600,000 tonnes of standard coal units to over 13 million tonnes of standard coal units.

In terms of policies to provide incentives and encouragement, initiatives related to energy prices have created reform in domestic mechanisms. China has carried out reform by taxing oil products, differentiating electricity prices in the high energy-consuming industries, introducing punitive electricity prices for ultra-high-energy products, and imposing taxation based on calculations of heat supply. Special funding was given for low energy use and reduced emissions, which amounted to CNY 225 billion worth of investment within five years. This was focused on making technology more energy efficient and promoting low-energy products, and it created a potential energy saving of 340 million tonnes of coal standard hours. China has also dealt with the taxation system, having initiated fiscal reform in respect of natural resources—including constant improvement of the tax rebate system for exports, revision of the policy on vehicle purchase tax and reform of taxation on vehicles and sea vessels—and announced fiscal policies offering tax breaks for low energy use, low water use and the multipurpose utilization of natural resources.

Following efforts made in various areas, by 2010 China's GDP per unit of energy consumed had fallen by 19.1% since 2005, which is equivalent to a reduction of over 1.46 billion tonnes of carbon dioxide emissions. During the period 2006~2010, China benefited from a mean annual increase of 6.6% in energy consumption to boost its economy's mean annual growth of 11.2%, while its energy coefficient fell from 1.04 during the 10th Five-Year Plan (2001~2005) to 0.59.

In terms of the development of new energy sources, China has vigorously developed natural gas and developed the use of unconventional oil and gas resources—for example, by promoting coalbed methane and shale gas. It has used policies such as financial subsidies, tax breaks and preferential electricity prices to determine and implement general strategies for controlling and using coal mine gas; it has vigorously advocated coal purification, and pioneered and encouraged the use of coal mine gas as well as the development of surface coalbed methane. As a result of these strategies, Chinese production of natural gas increased from 49.3 billion cu m in 2005 to 94.8 cu m in 2010, a mean annual increase of 14%, while natural gas accounted for 4.3% of total energy consumption by 2010. Altogether, 30.55 billion cu m of coalbed methane were extracted, of which 11.45 billion cu m were utilized. If converted into a reduction in emissions, this equates to a decrease in emissions of 170 million tonnes of carbon dioxide as a result of China's development of new energy sources.

Leaving aside the development of new fossil fuels, China has also given ample consideration to non-fossil fuels. For the duration of its 11th Five-Year Plan, it relied on national policy and financial investment to bolster the development and use of low carbon variants such as hydro and nuclear energy. By the end of 2010, the capacity of China's hydroelectric installations was 213 million kW, double the amount in 2005; it had a capacity of 10.82 million kW of nuclear facilities, with another 30.97 million kW under construction. As for other new types of energy, China's wind-electricity installation capacity increased from 1.26 million kW in 2005 to 31.07 million kW in 2010; photovoltaic electricity generators increased their output from 100,000 kW to 600,000 kW; solar-powered water heating installations covered a total of 168 million sq m; biomass electricity generation accounted for approximately 5 million kW; annual methane use was approximately 14 billion cu m, methane being used in approximately 40 million households across China; ethanol biofuel use was 1.8 tonnes; and the total cumulative contribution of all biofuels came to approximately 15 million tonnes of standard coal units.

Aside from controlling emissions of greenhouse gas during energy production, China has also introduced more stringent controls on greenhouse gas emissions in non-energy-consuming sectors, including the industrial production process, agricultural activities and the management of waste products. More precisely, it has replaced technology with raw materials such as cement clinker that is produced by replacing limestone with acetylene sludge; it has used technical processes to produce cement involving the addition of composite materials such as blast-furnace slag and fly ash; it has employed second and third level treatment processes to treat nitrous oxide emissions from the production of nitric acid, catalytic decompositions and thermo-oxidizing decompositions to treat nitrous oxide emissions from the production of nitric acid, and thermo-oxidation techniques to capture and disperse HFC-23. Besides, China is still accelerating the transformation of production methods of its livestock industry, and reducing emissions of methane and nitrous oxide from crop sowing and livestock rearing. In addition, it has initiated the implementation of projects using organic soil which carry increased subsidies. These now cover an area of nearly 30 million mu and include the use of technological measures such as replacing straw in the fields, growing crops using green manure and increasing the use of organic fertilizers. In urban areas, meanwhile, China has further improved its standards on urban waste, having initiated taxation on household waste management in a small number of places. It has promoted the use of advanced waste-burning technology, and created policies offering incentives for recycling landfill gases.

An official Chinese white paper states: "By the end of 2010, emissions of nitrous oxide from industrial production processes had basically stabilized at the level of 2005, while the acceleration of methane emissions has been brought under a measure of control."

In the five years from 2006 to 2010, another project undertaken by China to reduce emissions was to expand the carbon sink of its forests. This work consisted first of continuing existing ecological construction projects: a forest protection project in the "San Bei" area (north China, northwest China and northeast China); a forest protection project in the middle and lower reaches of the Yangtze river; a reforestation of farmland project; a protection of natural forests project; a Beijing-Tianjin sandstorm control project, and others. Later it developed an afforestation pilot project to create new carbon sinks, strengthened forest management and sustainable management methods, and increased the number of forests being saved and preserved. Additionally, it raised the standard level of investment available to subsidize afforestation in China's central fiscal budget from CNY 100 per mu to CNY 200 per mu, and it founded the China Green Carbon Foundation.

Data from 2010 shows that China has retained 62 million hectares of planted forest, and that it has 195 million hectares of forest nationwide. Total forest coverage has increased from 18.21% in 2005 to 20.36% in 2010, and 13.721 billion cu m of forest have been saved and preserved. There remain a total of 7.811 billion tonnes of carbon reserves in vegetation and forests in China. Similarly, the Chinese government has striven to increase carbon sinks on agricultural plots and grassed areas. Methods of protecting the grasslands have been adopted, including balancing the division of grassland and grazing land, the creation of non-grazing areas and temporary non-grazing areas, and the division of grasslands into rotational grazing areas. In addition, the stocking areas of the grasslands have been controlled to check their deterioration. The scope of the project to return grazing land to grassland has been extended, and construction of artificial forage grass areas and irrigation fields has been stepped up. Disaster prevention measures have been bolstered, while the area of the grasslands has been increased and the amount of carbon sinks there has been increased. By 2010, the area farmed using protection techniques covered 64.75 mu, whereas mechanized sowing was practiced over 167 million mu and mechanized threshing techniques were carried out over an area of 428 million mu.

During the five-year planning period from 2006 to 2010, China drew up or revised laws in this field including "Law on Renewable Energy", "Law for the Promotion of Circular Economics", "Law on Energy Conservation", "Law for the Promotion of Clean Production", "Law for the Maintenance of Land and Sea" and "Law for the Maintenance of Islands". It released "Regulations for Energy Saving in Private Housing", "Regulations on Energy Conservation in Public Organizations", "Regulations on Drought Resistance", and regulatory ordinances including "Provisional Solutions for the Appraisal and Investigation of Energy Conservation in Fixed Asset Investments", "Supervision and Management Techniques for Energy Conservation in High Energy Consuming Special Equipment" and "Provisional Solutions for the Supervision and Management of Energy Conservation and Emissions Reduction in National Enterprises". It also began to develop the initial stage of research into legislation on climate change.

Simultaneously, China established and implemented its "Chinese National Plan to Counteract Climate Change". This plan clarified the main areas and key tasks involved in work to counteract climate change. As required by the plan, each of the 31 provinces in China (and autonomous regions and directly governing cities except Taiwan, Hongkong and Macao) has already drawn up and completed local plans on climate change. They are all at the organizing and implementation stage. The work to counteract climate change has gradually been incorporated into each area's overall framework for economic and social development, and has become an integral part of the daily agenda of government officials at every level.

Will China's Energy Resource Tax Policy Work?

As Chapter 3 shows, facing such a global problem, China has already did a lot in the past decade, but for the new fiscal policy such as energy resource tax and carbon tax they had and will execute in the future, the effectiveness of these polices should still be discussed. Although as Chapter 2 shows, there were already a lot of studies in China to analysis the possible effect brought by carbon tax or energy resource tax, but these studies had missed an important point. China is a very large country. Because of its large population and area, every policy that is instituted has different effects on different regions and industries. These differences can be ascertained using any single-country model. In this study, a Multi-Regional CGE model (MRCGE hereinafter) is used to evaluate the possible regional effects of an energy resource policy in China.

4.1 Model Structure

The single-country, MRCGE used for this study is a static CGE model incorporating the assumption of a perfectly competitive market and minimization production cost for producers. For final users, all regional final users' saving would re-investment into one region's regional economy. International trade follows a small-country assumption and the Armington assumption. CO_2 emission in this model is assumed majorly from energy resource consumption and in this MRCGE model, the total CO_2 emission of China was treated as exogenous parameter which referred to the World Bank database.[①]

---

① This data could be find from the World Bank data base, see details from: http://data.worldbank.org/indicator/ EN. ATM. CO2E.PC?page=2

This model is based on the one-country static CGE model presented by Hosoe et al. (2004). Details related to model structure can be found in a discussion paper by Pu (2011). The model has five nests in its structure: production nest, household consumption nest, government activities nest, exports, and imports. The nests are as shown below.

In Fig. 4.1, showing the production structure, *VA* stands for the value-added composite, which takes the labor and capital for the CES function into account. The composite intermediate input is a composite of the same intermediate inputs of different regions. The output of industry $j(=1,\cdots,J)$ of region $r(=1,\cdots,R)$ is regarded as the composite of $VA_{s,j}$ goods and all composite intermediate inputs under the CES function.

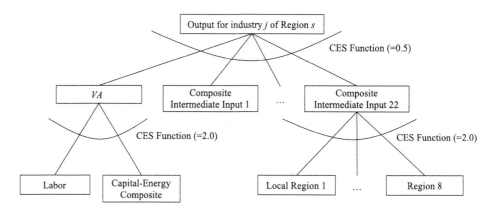

Fig. 4.1 Production Structure

In the following, Eq. (1)~Eq. (8) describe all production activities that include all CES functions presented in the figures. As explained in this paper, subscripts r and $s(=1,\cdots,S)$ stand for regional divisions, and $i(=1,\cdots,I)$ and j are divisions of industry or commodities.

$$VA_{s,j} = \alpha VA_{s,j} \left(\beta L_{s,j} L_{s,j}^{\rho_1} + \beta K_{s,j} K_{s,j}^{\rho_1} \right)^{\frac{1}{\rho_1}} \quad (1)$$

$$K_{s,j} = \left(\frac{PVA_{s,j} \alpha VA_{s,j}^{\rho_1} \beta K_{s,j}}{PK_{s,j}} \right)^{\frac{1}{1-\rho_1}} VA_{s,j} \quad (2)$$

$$L_{s,j} = \left(\frac{PVA_{s,j} \alpha VA_{s,j}^{\rho_1} \beta L_{s,j}}{PL_{s,j}} \right)^{\frac{1}{1-\rho_1}} VA_{s,j} \qquad (3)$$

$$X_{i,s,j} = \alpha X_{i,s,j} \left(\sum_{r \in R} \beta X_{r,i,j,s} XX_{r,i,j,s}^{\rho_2} \right)^{\frac{1}{\rho_2}} \qquad (4)$$

$$XX_{r,i,j,s} = \left(\frac{PX_{i,s,j} \alpha X_{i,s,j}^{\rho_2} \beta X_{r,i,s,j}}{PQ_{r,j}} \right)^{\frac{1}{1-\rho_2}} X_{i,s,j} \qquad (5)$$

$$VA_{s,j} = AVA_{s,j} Z_{s,j} \qquad (6)$$

$$X_{i,s,j} = AX_{i,s,j} Z_{s,j} \qquad (7)$$

$$PZ_{s,j} = PVA_{s,j} AVA_{s,j} + \sum_{i \in I} PX_{i,s,j} AX_{i,s,j} \qquad (8)$$

In the equations presented above, $K_{s,j}$ and $L_{s,j}$ respectively stand for capital input and labor input; $VA_{s,j}$ signifies a composite good of labor, energy and capital; also, $PVA_{s,j}$ is the price for it (every p variable is the price variable for the following explanation). Furthermore, $Z_{s,j}$ denotes output from the s region's j industry; $XX_{r,i,j,s}$ denotes the intermediate input in different regions; and $X_{i,s,j}$ signifies the composite intermediate input.

Fig. 4.2 portrays the structure of household consumption. In this structure, composite household consumption comprises household consumption of the same industries of different regions. This composite was also finished under a CES function. These activities are defined by Eq. (9)~Eq. (11).

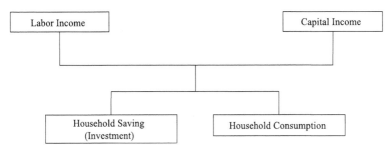

Fig. 4.2 Household Activities

$$XH_{i,s} = \alpha XH_{i,s} \left(\sum_{r \in R} \beta XXH_{r,i,s} XXH_{r,i,s}^{\rho_3} \right)^{\frac{1}{\rho_3}} \qquad (9)$$

$$XXH_{r,i,s} = \left(\frac{PXH_{i,s} \alpha XH_{i,s}^{\rho_3}}{PQ_{r,i}} \right)^{\frac{1}{1-\rho_3}} XH_{i,s} \qquad (10)$$

$$XH_{i,s} = \frac{\beta XH_{i,s}}{PXH_{i,s}} \left(PL_s FL_s + PK_s FK_s - SP_s - TD_s \right) \qquad (11)$$

The variable meanings for the equations above are the following: $XXH_{r,i,s}$ signifies household consumption of i good from region r consumed by region s consumers, $XH_{i,s}$ denotes composite household consumption, FK_s and FL_s respectively represent the factor endowments of capital and labor, SP_s stands for the total saving of private department, and TD_s represents the direct tax.

Fig. 4.3 portrays government activities. All government activities are assumed to be executed by the region's government. No central government exists. Local governments are distinguished in the model. As the structure shows, the government's income is based on taxation of three kinds: a production tax, energy tax, and direct tax. Direct taxes include labor income tax and capital tax. The government collects these taxes as government income and spends them on consumption and investment. Government activities are defined by Eq. (12)~Eq. (15). Eq. (16)~Eq. (18) determine the investment activities.

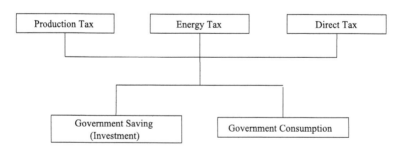

Fig. 4.3　Government Activities

$$XG_{r,i,s} = \frac{\beta G_{r,i,s}}{PQ_{r,i}} \left(\sum_{j \in J} TP_{s,j} + TD_s + TE_s - SG_s \right) \qquad (12)$$

$$TD_s = TDR \left(PK_s FK_s + PL_s FL_s \right) \qquad (13)$$

$$TP_{s,j} = TPR_{s,j} PZ_{s,j} Z_{s,j} \tag{14}$$

$$TE_s = \sum_{r \in R} \sum_{i \in I} \sum_{j \in J} TER_{r,i} XX_{r,i,s,j} + \sum_{r \in R} \sum_{i \in I} TER_{r,i} \left(XXH_{r,i,s} + INV_{r,i,s} + XG_{r,i,s} \right) \tag{15}$$

Among the equations presented above, Eq. (12) shows the relations among government consumption $XG_{r,i,s}$, production tax $TP_{s,j}$, direct tax TD_s, energy tax TE_s and government saving SG_s. Eq. (13)~Eq. (15) represent relations between each tax and the respective tax rate (TDR as the direct tax rate, $TPR_{s,j}$ as the production tax rate, $TER_{r,i}$ for the energy tax rate) and the tax base.

$$INV_{r,i,s} = \frac{\beta INV_{r,i,s}}{PQ_{r,i}} \left(SP_s + EXR \cdot FS_s + SG_s \right) \tag{16}$$

$$SP_s = SPR_s \left(PL_s FL_s + PK_s FK_s \right) \tag{17}$$

$$SG_s = SGR_s \left(TD_s + \sum_{i \in J} TP_{s,j} + \sum_{j \in J} TE_s \right) \tag{18}$$

Eq. (16)~Eq. (18) describe the relation between investment and saving. In Eq. (16), investment $INV_{r,i,s}$ is described as being related to private savings SP_s, the rate of currency exchange EXR, foreign savings FS_s, and government savings SG_s. Eq. (17) and Eq. (18) describe the quantitative relation between savings and the savings rate, such as private savings and private saving rate SPR_s and government savings and the government saving rate SGR_s.

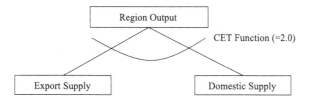

Fig. 4.4 Export Structure

In the export structure, the local total output is divided into export supply and domestic supply. This division procedure is based on a CET function. Eq. (19)~Eq. (21) describe the CET relation. In this CET expression, $Z_{r,i}$ stands for the total output, $E_{r,i}$ signifies the export supply, and $D_{r,i}$ represents the domestic supply.

$$Z_{r,i} = \alpha Z_{r,i} \left(\beta E_{r,i} E_{r,i}^{\rho_4} + \beta D1_{r,i} D_{r,i}^{\rho_4} \right)^{\frac{1}{\rho_4}} \tag{19}$$

$$E_{r,i} = \left(\frac{(1+TPR_{r,i})PZ_{r,i}\alpha Z_{r,i}^{\rho 4} \beta E_{r,i}}{PE_i} \right)^{\frac{1}{1-\rho 4}} Z_{r,i} \quad (20)$$

$$D_{r,i} = \left(\frac{(1+TPR_{r,i})PZ_{r,i}\alpha Z_{r,i}^{\rho 4} \beta E_{r,i}}{PD_i} \right)^{\frac{1}{1-\rho 4}} Z_{r,i} \quad (21)$$

Fig. 4.5 presents the import structure of the model. It might be said that the imported goods from the world market are combined with the local supply in a CES function under the Armington assumption. Those composite commodities are used to satisfy different demands, such as production input or household consumption in the local region.

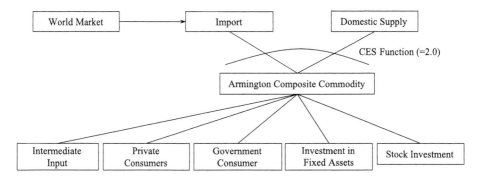

Fig. 4.5 Import Structure

$$Q_{r,i} = \alpha Q_{r,i} \left(\beta M_{r,i} M_{r,i}^{\rho 5} + \beta D2_{r,i} D_{r,i}^{\rho 5} \right)^{\frac{1}{\rho 5}} \quad (22)$$

$$M_{r,i} = \left(\frac{PQ_{r,i}\alpha Q_{r,i}^{\rho 5} \beta M_{r,i}}{PE_i} \right)^{\frac{1}{1-\rho 5}} Q_{r,i} \quad (23)$$

$$D_{r,i} = \left(\frac{PQ_{r,i}\alpha Q_{r,i}^{\rho 5} \beta D2_{r,i}}{PD_i} \right)^{\frac{1}{1-\rho 5}} Q_{r,i} \quad (24)$$

$$PE_i = EXR \cdot PWE_i \quad (25)$$

$$PM_i = EXR \cdot PWM_i \quad (26)$$

$$\sum_{r \in R} \sum_{i \in I} PWE_i E_{r,i} + \sum_{r \in R} FS_r = \sum_{r \in R} \sum_{i \in I} PWM_i M_{r,i} \quad (27)$$

Eq. (22)~Eq. (24) constitute the expression for the CES function, where $Q_{r,i}$ represents the Armington composite commodity. Eq. (25) shows the relation between the export price PE_i and the world export price PWE_i. Eq. (26) shows the relation between the import price PM_i and the world import price PWM_i. The last of these, Eq. (27), expresses the relations between exports, foreign savings, and imports.

$$Q_{r,i} = \sum_{s \in S} XXH_{r,i,s} + \sum_{s \in S} XG_{r,i,s} + \sum_{s \in S} \sum_{j \in J} XX_{r,i,s,j} + \sum_{s \in S} INV_{r,i,s} \qquad (28)$$

$$\sum_{j \in J} L_{s,j} = FL_s \qquad (29)$$

$$\sum_{j \in J} K_{s,j} = FK_s \qquad (30)$$

$$EX_r = \sum_{i \in I} \sum_{s \in S} \sum_{j \in J} XX_{r,i,s,j} + \sum_{i \in I} \sum_{s \in S} XH_{r,i,s} + \sum_{i \in I} \sum_{s \in S} XG_{r,i,s} + \sum_{i \in I} \sum_{s \in S} INV_{r,i,s} \qquad (31)$$

$$IM_s = \sum_{r \in R} \sum_{i \in I} \sum_{j \in J} XX_{r,i,s,j} + \sum_{r \in R} \sum_{i \in I} XH_{r,i,s} + \sum_{r \in R} \sum_{i \in I} XG_{r,i,s} + \sum_{r \in R} \sum_{i \in I} INV_{r,i,s} \qquad (32)$$

$$TR_s = IM_s - EX_r \qquad (33)$$

$$\sum_{s \in S} TR_s = 0 \qquad (34)$$

$$UU_s = \prod_{s \in S} XH_{i,s} \qquad (35)$$

$$SW = \sum_{s \in S} UU_s \qquad (36)$$

Eq. (28)~Eq. (36) represent the market-clearing condition of the model. Eq. (28) portrays the commodity market clearing condition. Eq. (29) and Eq. (30) represent the balance of the labor market and capital market. In the capital market, the model is based on the assumption that all capital in the country can be transferred freely among regions and industries. Eq. (31)~Eq. (34) represent the domestic trade situation in the model. In these equations, EX_r in Eq. (31) portrays one region's net export to other local regions, whereas IM_s in Eq. (32) represents one region's total import from other local regions. In Eq. (34), TR_s represents the net import from other local regions, and Eq. (35) represents the internal trade clean condition. Eq. (35) shows how household utility UU_s is calculated for these analyses. Eq. (36) is the objective function of the whole model, which describes the ultimate goal of this project: social welfare SW.

For this MRCGE model, the neoclassical closure was selected as the Marco-Closures for such CGE model. With this closure selection, the investment level was treated as endogenous parameter. This means, in this model, the market cleaning of investment market had been give up, investment level will be determined by saving level for each modeling region.

4.2 Data and Scenario

4.2.1 Regional Division

In this study, 31 mainland China provinces and municipalities were divided into eight regions based on geography and economic facts.

The regional division codes used are as shown in Table 4.1; the region-code relation can be referred simply from Picture 4.1.

Table 4.1   Regional Division Code(Taiwan, Hongkong and Macao excluded)

Code	Region	Included Provinces and Cities
A	Northeast	Heilongjiang, Jilin, Liaoning
B	North Municipalities	Beijing, Tianjin
C	Northern Coast	Hebei, Shandong
D	Eastern Coast	Jiangsu, Shanghai, Zhejiang
E	Southern Coast	Fujian, Guangdong, Hainan
F	Central	Shanxi, Henan, Anhui, Hubei, Hunan, Jiangxi
G	Northwest	Inner Mongolia, Shanxi, Ningxia, Gansu, Qinghai, Xinjiang
H	Southwest	Sichuan, Chongqing, Yunnan, Guizhou, Guangxi, Tibet

Picture 4.1   Region Code for 8 Division Regions in China

This manner of regional division is based not only on geography, but also on economics. Under this regional division, the eight divided regions have distinct economic characteristics. The Northeast region has long been an area of heavy industrial concentration. Its abundant mineral resources support and underpin heavy industry. The nation's largest oil field—Daqing—is located in the region, which also has a highly developed oil industry. In addition to its industrial advantages, the area is well known as a crop production base.

The North Municipalities region is special among the regional divisions. The region includes only two cities: Beijing and Tianjin. Although they might appear similar in area or in economic scale, the political significance of the two cities and the high-technology equipment manufacturing and financial services based there give sufficient reason for them to be regarded as independent area divisions.

The Northern Coast area has rich natural resources and various industries, such as manufacturing, energy, steel, petrochemicals, and high-technology industries. At the same time, the region has a rich output of agricultural products, such as cotton, edible oils, aquatic products, and vegetables. Its balanced economic structure gives this region strong competitive ability among regional economies.

The Eastern Coast and Southern Coast share certain similarities in their economic structures. Both areas have an export-oriented structure, with textile products and toys being the Eastern Coast's main export products and textiles and light chemical products being the Southern Coast's main export products. Benefiting from globalization, the two regions in such an export-oriented economic model have rapidly accumulated large amounts of wealth and have become the most economically developed areas in China.

The Central area includes six provinces that even ancient Chinese generals designated as the Central Plains area. This is a less well-off area: the major supplier of labor in China. Every year, millions of workers move from this area to coastal parts of the nation to find job opportunities. They are the major population of migrant workers nationwide. Aside from the labor supply, Shanxi province in this area is also the major supplier of coal for the nation.

The Northwest area includes several of the least-developed provinces in China, such as Ningxia, Gansu, and Qinghai. Although it might be the least developed region in the country, this region has many untapped mineral resources. These resource reserves endow this area with great economic development potential for the future.

The Southwest area is the last region in this division. The region in old China's strategic planning was called "the third line". This area has a complete industrial system and could achieve self-sufficiency. However, the industrial structure of this region has emphasized military-industrial production; this condition at some level has limited the economic development of the whole area. In addition to its industrial system, this area is known for its reserves of natural gas and rare earth minerals.

4.2.2 Industry Classification

Data used for the multi-region CGE model were based on China's 2000 multi-regional input-output matrix. The input-output matrix includes 8 regions and 30 commodity sectors. However, the present study was undertaken not only to investigate the effect of energy tax policy but also to observe whether the country's policy will affect the world in the future.

For the commodities, we referred to the GTAP7 database and reclassified the industry data sources. As Table 4.2 shows, the commodities were reclassified into 24 sectors.

Table 4.2 Reclassified Commodity Sectors

No.	Reclassified Commodity Sectors	China Multi-Regional IO 30 Sectors
1	Agriculture	Agriculture
2	Coal Mining	Coal Mining
3	Oil and Gas Mining	Oil and Gas Mining
4	Other Mining	Metal or Mining, Non-Metal or Mining
5	Food Manufacturing	Food Manufacturing and Tobacco Processing
6	Textile	Textile
7	Wearing Apparels	Wearing Apparel, Leather, Furs, Down and Related Products
8	Sawmills and Wood Products	Sawmills and Furniture
9	Paper Products	Paper and Products, Printing and Recording Medium Reproduction
10	Petroleum Processing and Coking	Petroleum Processing and Coking
11	Chemical Industry	Chemical Industry
12	Non-metallic Mineral Products	Non-metallic Mineral Products
13	Metal Smelting and Pressing	Metal Smelting and Pressing
14	Metal Products	Metal Products

		Continued
No.	Reclassified Commodity Sectors	China Multi-Regional IO 30 Sectors
15	Machinery Industry	Machinery Industry
16	Transport Equipment	Transport Equipment
17	Electrical Machinery and Equipment	Electrical, Machinery and Equipment
18	Electronic and Communication Equipment	Electronic and Communication Equipment Manufacturing
19	Other Manufacturing Industries	Measuring Instruments and Office Machinery, Machinery and Equipment Repair, Other Manufacturing Industries, Waste Disposal
20	Electricity, Water and Gas Supply	Electricity, Steam, and Hot Water Production and Supply, Gas Production and Supply, Tap Water Production and Supply
21	Construction	Construction
22	Transportation and Warehousing	Transportation and Warehousing
23	Commercial	Wholesale and Retail Trade
24	Services	Services

4.2.3 Scenarios

The scenario setup was considered from two different angles. On the one hand, because China wished to change their energy resource policy from almost zero tax to a 5% *ad valorem* tax on all energy goods for controlling carbon emission and because the policy would have been the first pilot in the western area of China, we decided to observe the possible effect caused by this pilot policy and the effects of future policies that might be followed. On the other hand, China promised during the 2009 Copenhagen meeting a 40%~45% per GDP CO_2 emission reduction. For that reason, we also investigated the policy impacts of this emissions reduction commitment.

To achieve these goals, we followed China's 2011 western area test fiscal policy (5% energy tax for energy resources) and its promised aim, setting up six different scenarios to evaluate the effectiveness of China's energy tax policy. These six scenarios are shown in Table 4.3.

As Table 4.3 shows, Scenario 1 is the pilot energy policy executed in 2011; Scenario 2 is the simulation of what will happen when the pilot policy is extended to the whole nation. Scenario 3 and Scenario 4 were used to analyze the macroeconomic impact of tax policy under half of China's Copenhagen commitment was reached by energy resource tax. Scenario 5 and Scenario 6 were used to evaluate the effects if

China reaches the goal promised in Copenhagen, which is to reduce the CO_2 emissions per capita by 40%~45%.

Table 4.3 Scenario Setup

Scenario	Content
Scenario 1 (S1)	5.0% energy resource tax in the western area
Scenario 2 (S2)	5.0% energy resource tax nationwide
Scenario 3 (S3)	12.5% energy resource tax nationwide
Scenario 4 (S4)	19.0% energy resource tax nationwide
Scenario 5 (S5)	26.5% energy resource tax nationwide
Scenario 6 (S6)	30.5% energy resource tax nationwide

To analyze the policy effectiveness, several indicators were chosen as analysis variables. These indicators are divisible into national level indicators, regional level indicators, and industry level indicators. National level indicators include national GDP losses, MAC, national CO_2 reduction, energy-intensive rate of change compared with the base scenario and several other macroeconomic indicators. Regional level indicators include reduction of petroleum and natural gas mining output, reduction of coal mining industry output, regional capital price rate of change, regional labor price rate of change, regional household utility (Equivalent Variation, EV) losses, regional per capita household utility losses and regional GDP rate of change. Additionally, we used the rate of change of different industries as the industry level indicator to observe the impact of different scenarios on the nation's industrial system.

4.3 Analysis Results

For this study, we used GAMS software to run all scenario simulations. For each scenario, the foreign exchange rate was set as numéraire, yielding the following results.

4.3.1 National Level Effect

Table 4.4 presents the key macroeconomic impacts and national GHG control impacts of China's energy resource tax changes based on the simulation of different tax-rate scenarios. The impact on real GDP is noticeable. The indicator RGDP

Reduction shows the real GDP decrease for each scenario. The decrease rate is enhanced with increased intensity of the levied tax rate.

Table 4.4 National Level Indexes

Index	S1	S2	S3	S4	S5	S6
CO_2 Emissions Reduction (Mt)	−65.31	−274.24	−650.23	−949.21	−1,269.86	−1,431.98
MAC (CNY/t)	26.75	44.79	50.33	54.98	60.24	63.01
Energy Intensity Reduction (%)	−0.18	−0.74	−1.74	−2.54	−3.38	−3.81
RGDP Reduction (Hundred Thousand CNY)	−17,470	−122,842	−327,240	−521,911	−765,012	−902,295
Household Consumption Rate of Change (%)	−0.07	−0.34	−0.84	−1.26	−1.75	−2.00
Government Consumption Rate of Change (%)	0.22	1.06	2.61	3.93	5.43	6.22
Investment Rate of Change (%)	0.01	0.05	0.13	0.19	0.25	0.28
National Indirect Tax Reduction (%)	0.00	−0.05	−0.12	−0.18	−0.23	−0.26
Nation Total Export Rate of Change (%)	0.02	0.06	0.16	0.26	0.37	0.44
Nation Total Import Rate of Change (%)	0.02	0.06	0.15	0.23	0.34	0.40

For other indicators, the household consumption rate of change declines gradually from −0.07% in Scenario 1 to −2% in Scenario 6. At the same time, the government consumption rate of change and the investment rate of change show growth. Especially for the rate of change of government consumption, although the most important income for the public sector, the indirect tax, was shown to decrease through each scenario, government consumption showed rapid growth attributable to the levy of the new energy resource tax. In Scenario 5 and Scenario 6, for which China reached its Copenhagen commitment, government consumption increased by 5.43% and 6.22%. Compared with other indicators, the significant increase of government consumption might be caused by the model structure, which assumed that the local government would put all of its newly levied tax into consumption and investment. In addition to the macroeconomic indicators presented above, other indicators such as national total exports and national total imports, showed only slight fluctuations, even under the most intense simulated policy. Although not as much change was shown by most macroeconomic indicators, the amount of carbon dioxide emissions reduction was impressive.

The CO_2 reduction quantities were calculated indirectly from the national outputs of coal, petroleum, and natural gas. For the calculation, we used 166 CNY/t for the coal price and 1,150 CNY/t for the petroleum and natural gas composite price. These energy prices were taken from the data of the China Development and Reform Commission Consumer Division and the Brent crude oil price in 1997. The CO_2 emission factor was taken from the IPCC (2.4567 for standard coal equivalent). It was used to calculate China's total carbon emissions reduction for each of the six scenarios. Results show that each energy-resource tax-policy scenario led to a significant reduction in CO_2 emission. As Table 4.4 shows, for Scenarios 1~6 (S1~S6), the national total CO_2 reductions were 65.31Mt, 274.24Mt, 650.23Mt, 949.21Mt, 1,269.86Mt and 1,431.98Mt. Compared with the total CO_2 emission in the base year, the national CO_2 reduction ratios were 2.08% in S1, 8.73% in S2, 20.71% in S3, 30.24% in S4, 40.46% in S5 and 45.63% in S6. These results are shown in Fig. 4.6.

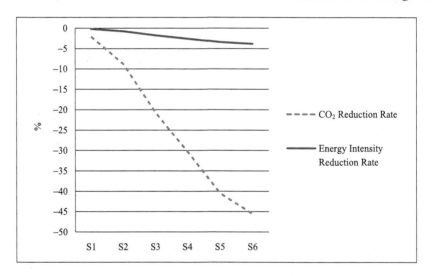

Fig. 4.6 CO_2 Reduction and Energy Intensity Rate of Change

The reduction ratio curve of China's energy intensity is shown in Fig. 4.6. For S1~S6, although the country's CO_2 emission decreased rapidly, the energy intensity did not. Even for S5 and S6, where China reached its Copenhagen commitment, the national energy intensity was reduced by only 3.38% and 3.81%. The large difference between these two indicators might be caused by their different base indexes. The base year's carbon dioxide emission could not be calculated directly using the benchmark solution of the MRCGE. Therefore, the number 3.13 Gt of CO_2 emission from Lee (2002) was used as the benchmark for comparison. However, for energy intensity, the benchmark comparison index was calculated from the model's benchmark solution. Perhaps for this reason, the two emission reduction efficiency indicators show such a dramatic difference.

Another important environment-related index, the marginal abatement cost (MAC), showed an increase while the policy strength increased gradually. For S1~S6, MAC was 26.75 CNY, 44.79 CNY, 50.33 CNY, 54.98 CNY, 60.24 CNY, and 63.01 CNY. As Fig. 4.7 shows, China's CO_2 emission MAC only increased rapidly from S1 to S2, where the energy resource tax was extended from a special region only to the whole nation. Subsequently MAC only showed moderate growth. Even under S6, the most powerful policy simulation scenario, MAC for CO_2 emission reduction by the energy resource tax was only 63.01 CNY, not even reaching the 8 USD under the base year's exchange rate. Compared with other studies using different policy simulations to analyze the reduction of China's carbon dioxide emission, such as those of He et al. (2002), Gao et al. (2004), and Wang et al. (2005), the MAC attributable to energy resource taxes is apparently much lower than it is for other policies.

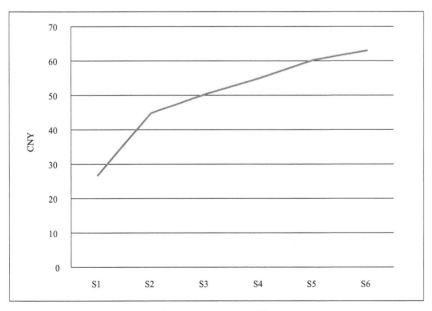

Fig. 4.7 National MAC under Different Scenarios

4.3.2 Regional Level Effect

Unlike national level indicators, regional level indicators were more focused on the regional differences in economic changes, abatement efficiency and fairness under different simulation scenarios. For differences in economic changes and regional GDP changes, regional household utility changes were used for analysis.

Regarding the GDP change figure, although the rate of change was small, five of eight regions showed a negative change in their GDP growth in each scenario, although the other three showed positive changes (the Central area in S1 also showed a negative GDP rate of change). Regional differences show that, for GDP change, areas based on heavy industry, such as the Northeast, Northern Coast and North Municipalities, and areas based on the chemical industry, such as the Eastern Coastland Southern Coast, show decreases in GDP that are explainable by the fact that these industries are highly energy-use-based industries. Areas that include important coal suppliers, such as Central and Northwest (each including one important coal supply province), or areas including natural gas resources, such as Southwest, show an increase in GDP. We assume that the simulated policy scenario also benefits activities in the energy sectors at some level.

Regarding household utility change, in most regions, it showed the same direction as the region's GDP change, but the Central, Northwest and Southwest regions showed conflicts between their GDP changes and household utility changes. This might be attributable to the CGE model itself. In the CGE model, one important assumption was that local government would take energy resource taxes in all simulation scenarios for government activities. That would show up as increases in government consumption and investment. On the one hand, during the calculation of regional GDP, such increases might have positive effects. These effects might engender increases, as are apparent in Fig. 4.8.

On the other hand, in addition to direct reduction of household income; the new energy resource tax will have the most negative effects on energy resource industries and energy-intensive industries. As described previously, major industries in which most people work in the Western and Central regions are such industries. For these reasons, although the growth in government activities might engender GDP growth in a region, the decrease in the region's personal income and the decrease in the region's major industry are expected to engender losses in the region's household utility.

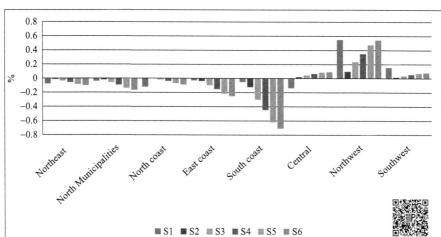

Fig. 4.8 Regional GDP Rate of Change

Unlike the regional household utility change, the per-capita household utility change presents a different picture of regional household utility losses from the perspective of fairness. As shown in Fig. 4.9, the Central area in each scenario suffered the most for its household utility, while the North Municipalities suffered the least. However, when the regional population was added to the analysis, a dramatic result was obtained, as shown in Fig. 4.10.

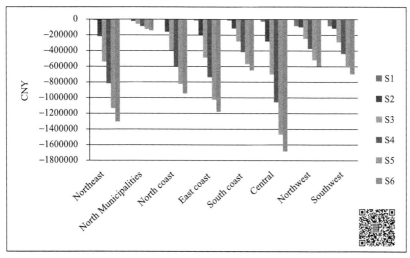

Fig. 4.9 Regional Household Utility Change

As shown in the bar graph above, the Central area was only the penultimate one for per capita household utility losses, while the North Municipalities became the third in most scenarios. This change might be caused by differences between regional populations. Surprisingly, the Northeast area had the most household utility loss through nationwide tax scenarios, the largest being 1238.3 CNY in S6. This very large amount of per capita household utility lost might be blamed on the Northeast area's industry structure. In the Northeast, the most affected industry was transport equipment with subsequent petroleum and natural gas, all energy intensive industries. As described in section 3, the major industries in which most people work in the Northeast region are industries of these two kinds. Just as for GDP and household utility change differences, an energy resource tax would engender a decrease in a region's personal income and in its major industry. Although the same thing happened in the Central and Western areas, it should be pointed out that the population in these two areas is almost three times that in the Northeast. Therefore, when per-capita household utility losses are considered, the Northeast showed the greatest loss: 1238.3 CNY per person under China's Copenhagen commitment scenario. In addition to the Northeast, Fig. 4.10 shows that, in terms of per capita household utility losses in China under carbon dioxide emission control fiscal policy, rich regions, such as Eastern Coast, Southern Coast, Northern Coast and two North Municipalities, Beijing

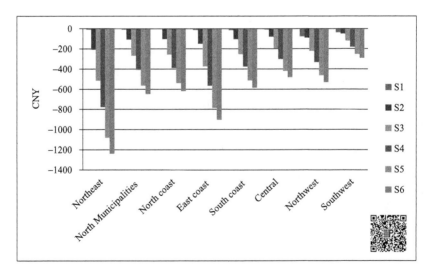

Fig. 4.10 Per Capita Household Utility Change

and Tianjin, would suffer more than the poor regions such as Central, Southwest and Northwest. Differences in regional household utility losses might seem unfair because the absolute numbers differ, but at some level, letting rich people pay more for environmental control might be fair in terms of social responsibility.

According to the regional petroleum and natural gas mining output changes shown in Fig. 4.11, every region showed a cut-back in petroleum and natural gas output in national level scenarios. In these reductions, the Southern Coast region suffers most, followed by the Central area in each national level scenario. As the figure shows, most regions had petroleum and natural gas mining industry output reductions of about 5%~10% under S5 and S6. For other national level scenarios, the reduction rate was about 0~5%. As the figure shows, the policy apparently has an especially large effect on the Southern Coast region. We checked the model results and found that the Southern Coast region was the only one where the output price of petroleum and natural gas mining industries changed by more than 8.87%. In other regions, the change was no more than 2%. It is noteworthy that such an impressive reduction in the Southern Coast's petroleum and natural gas mining output should have a strong correlation with the industries' output price change and might not be entirely the effect of the energy resource tax itself.

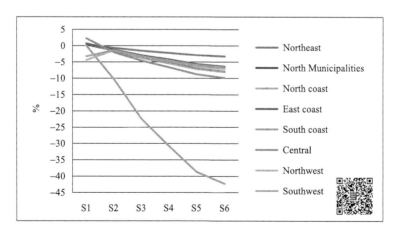

Fig. 4.11 Regional Petroleum and Natural Gas Mining Industry Output Reduction

Unlike petroleum and natural gas output change, the regional coal output change shown in Fig. 4.12 shows that under regional-level scenarios, regions other than the western areas would show increases in their coal mining industry output. However, when the fiscal policy is extended nationwide, all regions' coal mining industries show decreased output. Unlike the petroleum and natural gas mining industry, the reduction rate in the coal mining industry seems to fluctuate only slightly. This phenomenon might have an explanation. The Introduction described that China's main energy consumption was based on its self-produced raw coal resources. Therefore, the cardinal number of China's coal mining industry output was an extremely large number. It is precisely because of the very large reference value that the relative reduction in the amount of change is not might be readily apparent. It is noteworthy that the main coal mining areas in the Central and Northwest areas had the two highest rates of output reduction.

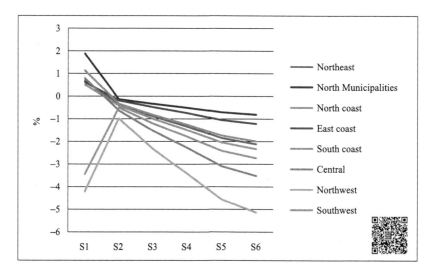

Fig. 4.12 Regional Coal Mining Industry Output Reduction

4.3.3 Industry Indicator

As shown in Fig. 4.13, under each scenario, most industries around the country showed decreased output. Of all industries, petroleum processing and the coking industry decreased the most (about 10%) under S6, followed by the oil and gas mining industry (9%), electricity, water and gas supply (3.5%) and the coal mining industry (about 3%) under the same scenario. It is noteworthy that these industries are all energy-intensive industries.

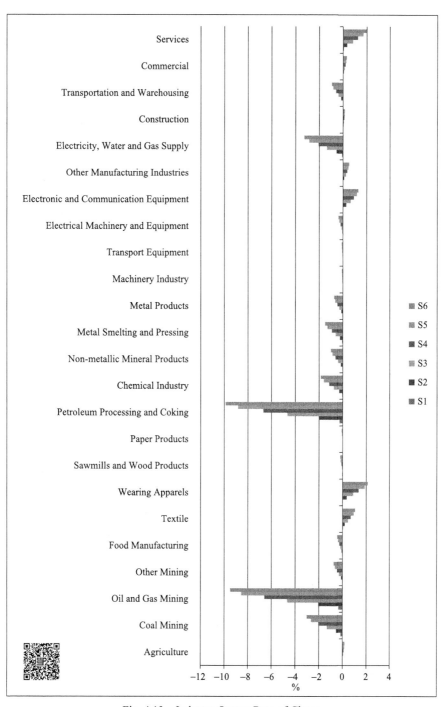

Fig. 4.13 Industry Output Rate of Change

The figure shows that regional policy as in S1 will not engender a large shock on the nation's industrial output. However, an energy resource tax levied at the national level strongly impacted the economy, especially energy-intensive industries.

Moreover, when most industries in all scenarios were in recession, services, commercial activities, construction, electrical machinery and equipment, wearing apparel and textiles showed increased business activity in all scenarios. These industries' output increase might have occurred mainly because their developments are less dependent on energy sectors, a key control parameter in the model.

Another interesting thing for industry level output rate of change is that it seems each industry has only one direction for its effect by such scenario policies. No matter such policy was regional only or national wide. But for different regions, industry output rate of change tells different stories. Take coal mining industry as example. In scenario 1, regions besides western provinces (northwest region and southwest region) show increase for their regional output, but when policy expands to national wide, coal mining industry for those regions also shows decrease. Same situation also happened in several other industries. But since some sector's major produce regions show significant decrease for industry output. Thus, for national level industry output rate of change, no matter for scenario which policy shock is regional only or for scenarios which policy shock is national wide, each industry had the same direction for their output rate of change.

4.4 A Dynamic Extension

4.4.1 Model Extension

The DMRCGE model used in this study is a multi-regional dynamic CGE model incorporating the assumption of perfectly competitive market and zero profit. International trade followed a small country assumption and Armington assumption. Substitution between capital and energy was considered during the production process. This model was based on the one-country static CGE model of Hosoe et al. (2004), details about static part of model structure could be referred to Pu and Hayashiyama (2012).

For dynamic part of this model, we had referred to Ban's (2007) multi region dynamic of Japan, and set capital's market interest rate at 5%, capital depreciation rate at 4% and increase rate of labor force at 13% for dynamic model simulation.

$$PK_{t+1} = P_t \tag{37}$$

$$P_{t-1} = (1+r)P_t = (1-\delta)P_t + RK_t \tag{38}$$

$$(n+\delta)K_t = I_t \tag{39}$$

$$RK_t K_t = VK_t \tag{40}$$

$$I_0 = \frac{n+\delta}{r+\delta}VK_0 \tag{41}$$

As for the model to reach dynamic equilibrium in every simulation year, condition from Eq. (37) to Eq. (41) should be satisfied. In these equations, P_t was commodity's market price level at t period, PK_{t+1} is the present value of capital's marginal productivity of each period. Moreover, RK_t was the rent of the leased capital, r is capital's market interest rate, n stands for increase rate of labor force and δ is the capital depreciation rate. Besides, K_t, I_t and VK_t are capital stock, investment and capital income for t period.

$$\frac{I_T}{I_{T-1}} = \frac{Y_T}{Y_{T-1}} \tag{42}$$

For the terminal condition setting, we followed the condition in the research of Rutherford and Paltsev (2002), made the growth rate of investment equal to the growth rate of income as Eq. (42) shows.

4.4.2 Data Resource and Scenario Setup

Same as the static model, data used for this DMRCGE model was based on 2000 China's multi-regional input-output matrix (2003). The input-output matrix includes 8 regions and 30 commodity sectors. The region division situation was as shown as Table 4.1.

Following this division, western area of China could be considered to include region northwest and southwest, but for energy resource production, it must be mentioned that main coal energy producers were present in central area and northwest. Besides above, following this region division, east coast and south coast region could be considered as two most outsourcing economy areas in China and northeast region's economy was heavily based on its heavy industry.

For 30 commodity sectors, we made a linkage between data in this study and GTAP data which mentioned by Burniaux and Truong (2002), reclassified industry classifications into 24 new sectors for 2000 China's multi-regional input-output matrix data sources. We made this new classification under two proposals. On one hand, we hope to make the model more concise through simplifying calculating process; on the other hand, we are going to make a connection between this DMRCGE model and the GTAP, and to make a linkage between this model and the GTAP-E model. If to do so, the first step to achieve this linkage was synchronizing two models' data sector. These 24 new sectors were: Agriculture, Coal Mining, Oil and Gas Mining, Other Mining, Food Manufacturing, Textile, Wearing Apparels, Sawmills and Wood Products, Paper Products, Petroleum Processing and Coking, Chemical Industry, Non-metallic Mineral Products, Metal Smelting and Pressing, Metal Products, Machinery Industry, Transport Equipment, Electrical Machinery and Equipment, Electronic and Communication Equipment, Other Manufacturing Industries, Electricity, Water and Gas Supply, Construction, Transportation and Warehousing, Commercial and Services.

From this research, we had selected the time period from 1997 to 2012 as the simulation time period and made a 5% western province only *ad valorem* energy resource tax been executed in at the year 2009 to compare the effects of such a GHG control policy on China's ecological system and economy. Simulation results were shown as follows.

4.4.3 Results

For dynamic simulation, MSPGE pack in GAMS software was used as computation tool for scenariosimulation.

Table 4.5 was the rate of change in terminal stage for each region's industry output compared with baseline solution. As the result shows, with the execution of *ad valorem* energy resource tax, most energy intensive industries in each region had been strongly affected by the policy. Industries which were traditionally known as heavy industries such as Non-metallic mineral products, Metal smelting and pressing, Metal products, Machinery industry and Transport equipment had been affected the most. At the same time, industries those were less relied on energies such as Textile and Wearing apparels had showed significant growth in most regions.

Table 4.5 Regional Industry Output Rate of Change in the Year 2012

Sector	Northeast	North Municipalities	North Coast	East Coast	South Coast	Central	Northwest	Southwest
Agriculture	−4.58	−2.53	−11.18	−6.25	−3.71	43.45	−3.02	−10.55
Coal Mining	1.82	−2.53	−8.04	−3.70	−3.98	−33.23	2.49	6.51
Oil and Gas Mining	3.83	−3.85	−2.89	2.11	−5.48	−27.36	−2.65	5.86
Other Mining	−2.95	−12.98	−19.54	−11.66	−9.53	−34.40	−5.11	1.80
Food Manufacturing	−2.76	−8.65	−14.84	−3.81	−2.78	29.63	1.75	−6.33
Textile	−0.89	38.88	12.91	31.31	18.75	28.17	6.66	5.43
Wearing Apparels	−4.02	−9.90	−17.86	0.29	15.46	9.37	6.67	22.89
Sawmills and Wood Products	−5.86	−11.24	−17.22	−10.11	−6.09	1.28	1.27	−8.79
Paper Products	−1.96	−1.53	−6.72	−4.77	0.53	6.66	1.78	−0.53
Petroleum Processing and Coking	−0.91	−4.79	−10.32	−3.80	−5.85	−18.10	−3.63	1.31
Chemical Industry	2.93	1.69	−4.36	4.29	5.14	−3.65	1.59	−1.37
Non-metallic Mineral Products	−5.63	−17.36	−33.81	−13.95	−11.53	−11.38	−9.79	−14.66
Metal Smelting and Pressing	−0.30	−11.12	−18.85	−10.23	−9.21	−29.22	−8.68	0.58
Metal Products	−3.65	−9.32	−28.83	−10.57	−7.94	−4.51	−11.63	−7.22
Machinery Industry	−6.81	−13.97	−28.63	−13.67	−11.24	−20.01	−7.55	−7.38
Transport Equipment	−3.38	−14.61	−23.21	−13.84	−10.26	−14.16	−3.23	2.17
Electrical Machinery and Equipment	−6.01	−14.08	−33.00	−12.55	−6.87	12.93	−7.49	−9.84
Electronic and Communication Equipment	13.58	12.07	−18.78	−5.62	8.55	−14.60	−4.03	−2.82
Other Manufacturing Industries	−2.63	−6.77	−9.98	−4.54	1.22	1.64	0.20	−3.20
Electricity, Water and Gas Supply	−2.69	−0.65	−11.16	−2.55	−4.35	−6.43	−0.85	1.69
Construction	−13.90	−20.51	−49.32	−20.52	−14.86	−11.18	−13.48	−21.49
Transportation and Warehousing	2.96	−4.16	−12.67	−1.99	−1.79	−13.31	1.84	8.91
Commercial	−2.74	−4.66	−3.89	0.28	0.22	−13.91	−1.60	−4.59
Services	−5.58	−5.98	−26.89	−9.22	−4.26	16.41	−2.11	−1.57

Besides industry effects, compared with Business as Usual (BAU) scenario, a GHG control policy will also significantly affect the macro economy of each region in very different directions.

As Fig. 4.14 shows, with the pilot energy resource tax been carried out, compared with BAU scenario, every region would show decrease in its total regional investment. For these eight regions, north coast had the largest decrease at the terminal stage followed by southeast area and east coast.

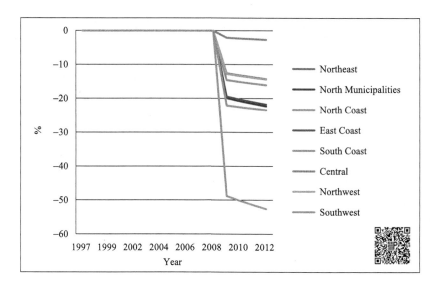

Fig. 4.14　Regional Investment Rate of Change

But for regional household consumption (Fig. 4.15), although there were seven regions showed decrease in their regional private consumption, only north coast which was one of the heavy industry based area had the most significant reduction, other six regions was not significantly reduced in their private consumption compared to the northern coast. Besides these seven regions, central area region had impressive growth for its household consumption.

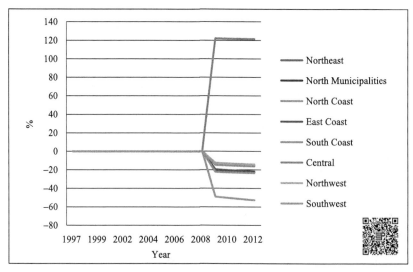

Fig. 4.15　Regional Household Consumption Rate of Change

As the Gross Regional Product (GRP) change for each region or the Gross Domestic Product (GDP) change for the whole nation present in Fig. 4.16 and Fig. 4.17, it could been seen that as the energy resource tax executed, China's national GDP would decrease up to 2.6% compared with BAU, but the regional GRP rate of change had a different story.

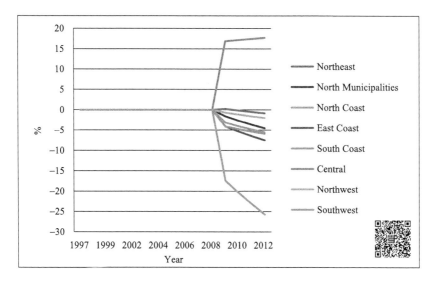

Fig. 4.16 Regional GRP Rate of Change

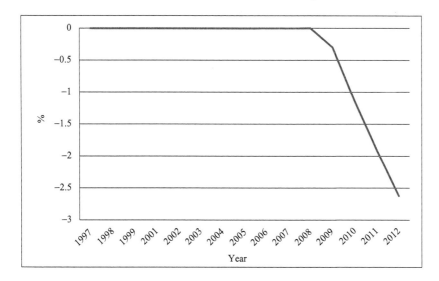

Fig. 4.17 National GDP Rate of Change

As Fig. 4.16 shows, 7 of 8 regions presented decrease in their GRP but central area of China had increase in its GRP from the very beginning of the pilot policy started. This trend was same as the trend in regional household consumption change. Further, if we took into account that the reduction in investment of central area and the reduction of the regional industrial output in central area were less than other regions, such an energy resource tax policy would benefit China's energy resource production area but had a negative impact on the overall macroeconomic.

To testify ecological effect of the scenario policy to the country, we incorporate energy resources as the major carbon emission factor of economic activities in this model. Energy production in this model is produced by the coal mining sector, petroleum and natural gas mining sector. China's energy self-sufficiency rate is higher than 95% for past three decades, and it can be assumed that China's energy consumption products are mainly produced by its own energy production sector. Thus, the more China's energy production sector's output decreased, the more carbon dioxide emission reduction will be achieved for the nation.

Under this setting, we chose regional coal mining industry's output rate of change compared with BAU, and regional oil and gas mining industry's output rate of change compared with BAU as indexes in this research to judge the ecological effect to different regions of China. As it represented in Fig. 4.18 and Fig. 4.19, this policy

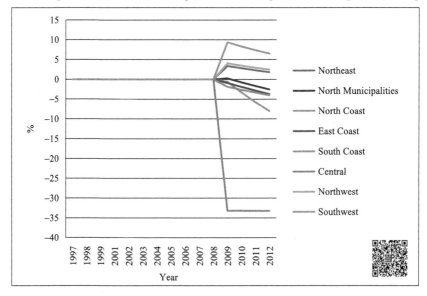

Fig. 4.18 Regional Coal Mining Output Rate of Change

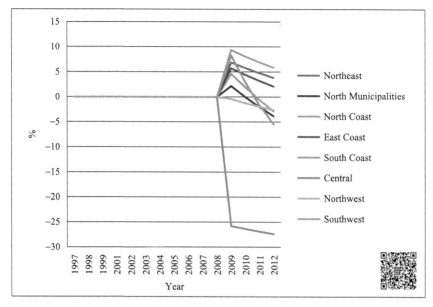

Fig. 4.19 Regional Oil and Gas Mining Output Rate of Change

made a significant control for the GHG emission of central area in China, meanwhile, a decrease of regional coal mining output was observed in other 4 regions beside central region at the beginning of the policy been executed. Although the oil and gas mining output in five areas was decreased at the terminal stage, such output in regions beside northwest and central was increased at the start of the policy. It was assumed that, the ecological effect for such an energy resource control policy may be more efficient for control of coal resource application than that of petroleum and natural gas resource.

4.5 Conclusion

A multi-regional CGE model was built to evaluate the effects of an energy resource tax on China's regional economy in this chapter, especially under its Copenhagen commitment. Following simulation results, two main conclusions were summarized as follows.

Firstly, a significant CO_2 reduction can be accomplished under the performance of energy resource tax in certain ranges while the minor adverse effects on macroeconomic of China will be made. During the scenario simulations, the nation's total CO_2 emission rate is decreased by 2.08% under a 5% energy resource tax executed in west China. When the policy extends to national wide, the total CO_2 emission rate can be decreased by 8.74%. Meanwhile, the MAC is 44.79 CNY and the country's RGDP has a 0.175% reduction compared with base year data. When a 12.5% *ad valorem* tax is executed, China can carry out a 20.71% reduction rate for its total GHG emission, this is about half of the promised goal which China had announced on COP15 in Copenhagen. While the MAC for this policy result is 50.33 CNY and the country's RGDP reduction rate is 0.46%. But if China wants to reach its 45% CO_2 emission reduction commitment only depending upon the energy resource tax policy, the *ad valorem* tax rate should set at 30.5%, the MAC at that time will be 61.01 CNY and China's national RGDP will reduce for 1.3%. In conclusion, application of appropriate energy resource tax to meet China's CO_2 reduction mission is feasible.

Secondly, as the industry structure and economic level of development is different through 8 regions; the execution of energy resource tax may lead to unfairness for regional per capita household utility losses. Thus, when setting up energy resource tax ratio, considered the industry structure differences and economic level of development differences between regions, China should follow the common but differentiated principle making the region's resources tax rate differences. At the same time, economically developed regions should take more emission reduction payments and use financial transfer payment to support less developed regions to reduce the loss.

Also, a dynamic CGE model proved that an *ad valorem* energy resource tax can reduce carbon dioxide emissions in main resource production region of China. However, this fiscal policy might cause different effects in different regions of China. Energy resource based regions might benefit from this policy, although energy-intensive industries and heavy-industry-based areas might be strongly affected by the energy resource tax. It is also concluded that the energy resource tax policy seems more efficient in application control of coal resource than that of petroleum and natural gas resource.

What Will Happen When the World Works Together? 5

In the previous chapter, a static MRCGE model was created to evaluate the effect of an energy resource tax on different regions in China. For such analysis, the CO_2 for base year simulation was exogenous and the analysis only considered economic and ecological effects in China. However, as discussed in Chapter 2, for a global issue such as GHG reduction, consideration of a single country may not be sufficient. Therefore, the MRCGE model was linked to the GTAP model to facilitate evaluation of the effect for worldwide execution of climate policy in different regions in China and throughout the world. We call this a MRCGE-GTAP model.

5.1 Model Structure

The MRCGE-GTAP model used for this study is also a static CGE model incorporating the assumption of a perfectly competitive market and minimization of production costs. International trade only follows the Armington assumption, as it does for the MRCGE model in Chapter 4. However, for the MRCGE-GTAP model, total CO_2 emission for each country (region) is endogenous. The situations for investment and model closure are both the same as for the MRCGE model in the previous chapter. Like the MRCGE model, this model is based on the one-country static CGE model presented by Hosoe, Gasawa, and Hashimoto (2004) and the static MRCGE model for China by Pu (2011). The model structure can be divided into two parts: a structure that describes activities in different regions of China (MRCGE) and a structure that describes activities throughout the world (GTAP). Fig. 5.1~Fig. 5.5 describe the MRCGE part of the model, which is the same as the model structure described in Chapter 4. Thus, the equations describing the MRCGE part of the model are not repeated here. Fig. 5.6~Fig. 5.10 describe the GTAP part of the model. Eq. (80)~Eq. (86) describe the market cleaning condition for the GTAP part of the model. Eq. (87)~Eq. (90) are the necessary conditions for establishing the linkage model.

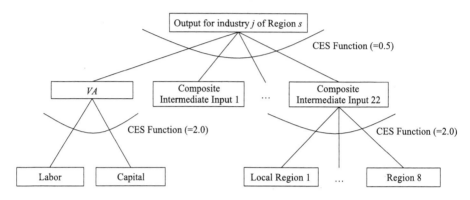

Fig. 5.1 Production Structure

As in Chapter 4, for the production structure shown in Fig. 5.1, VA denotes the value-added composite, which takes labor and capital for the CES function into account. The composite intermediate input combines the same intermediate inputs of different regions. The output for industry j in region r is regarded as the composite of $VA_{s,j}$ goods and all composite intermediate inputs under the CES function.

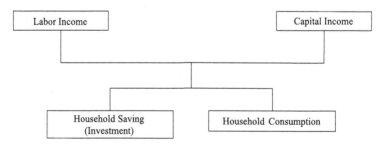

Fig. 5.2 Household Activities

Fig. 5.2 shows the composite household consumption structure, which comprises household consumption for the same industries for different regions according to a CES function.

What Will Happen When the World Works Together? 73

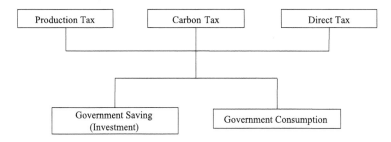

Fig. 5.3 Government Activities

Fig. 5.3 shows government activities for China in the model. It is assumed that all government activities are by the local regional government and no central government exists. Government income is derived from three types of taxation: production tax, energy tax, and direct tax. Direct taxes include labor income tax and capital tax. The government collects these taxes as government income and spends them on consumption and investment.

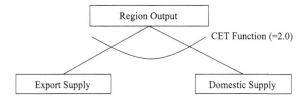

Fig. 5.4 Export Structure

In the export structure, the total local output is divided into export supply and domestic supply according to a CET function.

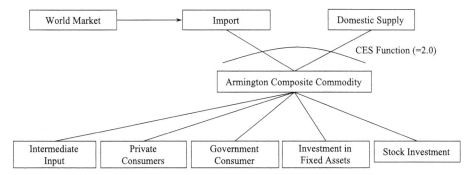

Fig. 5.5 Import Structure

Fig. 5.5 shows the import structure of the MRCGE part of the model. Goods imported from the world market are combined with the local supply in a CES function under the Armington assumption. Those composite commodities are used to satisfy different demands, such as production input or household consumption in the local region.

The world model used for this research has a simplified GTAP structure. For example, the original GTAP model includes a World Bank sector to describe capital flow throughout the world. In the model used here, this sector was omitted and cash flow is mainly described by changes in the general account for each region.

In the production structure shown in Fig. 5.6, WVA denotes the value-added composite, which takes labor, capital, land and natural resources for the CES function into account. The output by industry j in region rr is regarded as the composite of $WVA_{s,j}$ goods and all intermediate inputs under the CES function.

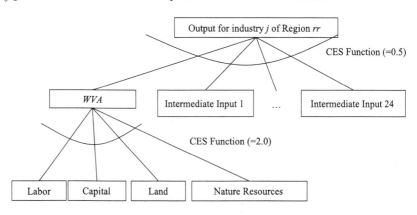

Fig. 5.6 Production Structure for World Region

Eq. (43)~Eq. (55) describe the production activities that include all the functions presented in the figures. Throughout this research, subscripts rr $(1,\cdots,RR)$ and oc $(=1,\cdots,OC)$ denote different regions, and i $(=1,\cdots,I)$ and j $(=1,\cdots,J)$ denote different industries or commodities.

$$WVA_{rr,i} = \alpha WVA_{rr,i} \left[\begin{array}{l} \beta WLAB_{rr,i}WL_{rr,i}^{\rho w1} + \beta WK_{rr,i}WK_{rr,i}^{\rho w1} + \\ \beta WLAN_{rr,j}WLAN_{rr,j}^{\rho w1} + \beta WENE_{rr,i}WENEe_{rr,i}^{\rho w1} \end{array} \right]^{\frac{1}{\rho w1}} \quad (43)$$

$$WLAB_{rr,i} = \left(\frac{PWVA_{rr,i} \alpha WVA_{rr,i}^{\rho_{w1}} \beta WLAB_{rr,i}}{(1+\tau WLAB_{rr,i})PWLAB_{rr}} \right)^{\frac{1}{1-\rho_{w1}}} WVA_{rr,i} \qquad (44)$$

$$WK_{rr,i} = \left(\frac{PWVA_{rr,i} \alpha WVA_{rr,i}^{\rho_{w1}} \beta WK_{rr,i}}{(1+\tau WK_{rr,i})PWK_{rr}} \right)^{\frac{1}{1-\rho_{w1}}} WVA_{rr,i} \qquad (45)$$

$$WLAN_{rr,i} = \left(\frac{PWVA_{rr,i} \alpha WVA_{rr,i}^{\rho_{w1}} \beta WLAN_{rr,i}}{(1+\tau WLAN_{rr,i})PWLAN_{rr}} \right)^{\frac{1}{1-\rho_{w1}}} WVA_{rr,i} \qquad (46)$$

$$WENE_{rr,i} = \left(\frac{PWVA_{rr,i} \alpha WVA_{rr,i}^{\rho_{w1}} \beta WENE_{rr,i}}{(1+\tau WENE_{rr,i})PWENE_{rr}} \right)^{\frac{1}{1-\rho_{w1}}} WVA_{rr,i} \qquad (47)$$

$$WVA_{rr,i} = AWVA_{rr,i} WZ_{rr,i} \qquad (48)$$

$$WX_{rr,i,j} = AWX_{rr,i,j} WZ_{rr,i} \qquad (49)$$

$$PWZ_{rr,i} = PWVA_{rr,i} AWVA_{rr,i} + \sum_{i \in I}(1+\tau WX_{rr,i,j})PWX_{rr,i,j} AWX_{rr,i,j} \qquad (50)$$

$$TWLAB_{rr,i} = \tau WLAB_{rr,i} PWLAB_{rr} WLAB_{rr,i} \qquad (51)$$

$$TWK_{rr,i} = \tau WK_{rr,i} PWK_{rr} WK_{rr,i} \qquad (52)$$

$$TWLAN_{rr,i} = \tau WLAN_{rr,i} PWLAN_{rr} WLAN_{rr,i} \qquad (53)$$

$$TWENE_{rr,i} = \tau WENE_{rr,i} PWENE_{rr} WENE_{rr,i} \qquad (54)$$

$$TWX_{rr,i,j} = \tau WX_{rr,i,j} PWX_{rr,i,j} WX_{rr,i,j} \qquad (55)$$

In the equations presented above, $WL_{rr,i}^{\rho_{w1}}$, $WK_{rr,i}^{\rho_{w1}}$, $WLAN_{rr,j}^{\rho_{w1}}$ and $WENEe_{rr,i}^{\rho_{w1}}$ denote for labor, capital, land and energy resource inputs, respectively, for different region in the GTAP part of the model. $WVA_{rr,i}$ is a composite good comprising labor, energy resources, land and capital in world regions and $PWVA_{rr,i}$ is its price for it (P denotes price for a variable hereafter). $WZ_{rr,i}$ denotes output from industry i in region rr. $WX_{rr,i,j}$ is the intermediate input in different regions. $TWLAB_{rr,i}$ is the tax for labor input (T denotes tax for a variable hereafter).

Fig. 5.7 portrays the structure of household activities for world region rr. In this structure, composite consumption comprises household consumption of the same industry for different regions. Besides consumption, as in the MRCGE part, households earn their income from primary factors they own (labor, land, capital and natural resources), and savings, which they can invest in the economic activity of their region. Eq. (56)~Eq. (58) define this activity.

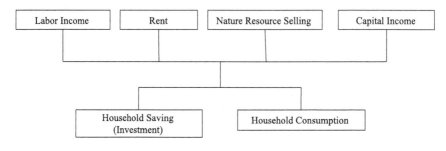

Fig. 5.7   Private Sector Activities for World Region

$$WXH_{rr,i} = \frac{\beta WH_{rr,i}}{(1+\tau WH_{rr,i})^{\sigma wh} PWQ_{rr,i}^{\sigma wh}} \cdot \left[\frac{\begin{bmatrix} PWLAB_{rr}FWLAB_{rr} + PWK_{rr}FWK_{rr} + \\ PWLAN_{rr}FWLAN_{rr} + PWENE_{rr}FWENE_{rr} - WHS_{rr} \end{bmatrix}^{(1-\sigma wh)}}{(\sum_{i \in I} \beta WH_{rr,i}(1+\tau WH_{rr,i})^{(1-\sigma wh)} PWQ_{rr,i}^{(1-\sigma wh)})} \right]$$

(56)

$$WHS_{rr} = \delta_{whs_{rr}} \begin{bmatrix} PWLAB_{rr}FWLAB_{rr} + PWK_{rr}FWK_{rr} + \\ PWLAN_{rr}FWLAN_{rr} + PWENE_{rr}FWENE_{rr} \end{bmatrix}$$

(57)

$$TWD_{rr} = \tau_{wtd_{rr}} \begin{bmatrix} PWLAB_{rr}FWLAB_{rr} + PWK_{rr}FWK_{rr} + \\ PWLAN_{rr}FWLAN_{rr} + PWENE_{rr}FWENE_{rr} \end{bmatrix}$$

(58)

$WXH_{rr,i}$ denotes household consumption of good i in region rr. $FWLAB_{rr}$, FWK_{rr}, $FWLAN_{rr}$ and $FWENE_{rr}$ represent factor endowments for capital, labor, land and natural resources, respectively. WHS_{rr} is total savings for the household sector, and TWD_{rr} represents direct tax.

Fig. 5.8 shows government activities for different world regions besides China. It is assumed that all government activities are executed by the national (regional) government. Each area has only one united government distinguished in the model. As the structure shows, government income is derived from several taxation streams: production tax, direct tax, consumption tax, tax from value-added inputs, tax from intermediate inputs, tax earned from exports and tax earned from tariffs. The government collects these taxes as income and spends them on consumption and investment (like households, in this model, we assume that a government uses all its savings to invest in the regional economy). Government activities are defined by Eq. (59)~Eq. (61). Eq. (62) and Eq. (63) describe investment activities.

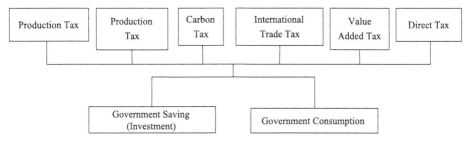

Fig. 5.8 Government Activities for World Region

$$WXG_{rr,i} = \frac{\beta XG_{rr,i}}{PWQ_{rr,i}{}^{\sigma XG}} \cdot \begin{bmatrix} \sum_{j \in J}(TWLAB_{rr,j} + TWK_{rr,j} + TEWLAN_{rr,i} + TWENE_{rr,i} + TWZ_{rr,i}) + \\ \sum_{i \in I}\sum_{j \in J} TWX_{rr,i,j} + TWD_{rr} + \sum_{i \in I}(TWH_{rr,i} + TEC_{rr,i} + TMC_{rr,i}) + \\ \sum_{i \in I}\sum_{oc \in OC}(TWE_{rr,i,oc} + TWM_{rr,i,oc}) - GWS_{rr} \end{bmatrix}$$

(59)

$$TWZ_{rr,i} = \tau WZ_{r,i} PWZ_{rr,i} WZ_{rr,i}$$

(60)

$$GWS_{rr} = \delta_{GWS_{rr}} \begin{bmatrix} \sum_{j \in J}(TWLAB_{rr,j} + TWK_{rr,j} + TEWLAN_{rr,i} + TWENE_{rr,i} + TWZ_{rr,i}) + \\ \sum_{i \in I}\sum_{j \in J} TWX_{rr,i,j} + TWD_{rr} + \sum_{i \in I}(TWH_{rr,i} + TEC_{rr,i} + TMC_{rr,i}) + \\ \sum_{i \in I}\sum_{oc \in OC}(TWE_{rr,i,oc} + TWM_{rr,i,oc}) \end{bmatrix}$$

(61)

Eq. (59) shows the relations among government consumption $WXG_{rr,i}$, each tax type and government saving GWS_{rr}. Eq. (60) describes the construct for the industry output tax and Eq. (61) expresses the relationship between government saving and all tax types.

$$WXI_{rr,i} = \frac{\beta WI_{rr,i}(HWS_{rr} + GWS_{rr} + \varepsilon_{rr} SWF_{rr})}{PWQ_{r,i}{}^{\sigma wi}\left(\sum_{i \in I} \beta WI_{rr,i} PWQ_{rr,i}{}^{(1-\sigma wi)}\right)}$$

(62)

$$SWF_{rr} = \begin{bmatrix} \sum_{i \in I} \sum_{oc \in OC} \left(PWMM_{rr,i,oc} MM_{rr,i,oc} - PWEE_{rr,i,oc} EE_{rr,i,oc} \right) + \\ \sum_{i \in I} \left(PWMC_{rr,i} MC_{rr,i} - PWEC_{rr,i} ME_{r,i} \right) \end{bmatrix} \quad (63)$$

Eq. (62) and Eq. (63) describe the relation between investment and savings. In Eq. (62), investment $WXI_{rr,i}$ is related to private savings HWS_{rr}, government savings GWS_{rr} and foreign savings SWF_{rr}, where ε_{rr} is the exchange rate in region rr. Eq. (63) describes the quantitative relation between foreign savings for a region and international trade.

In the export structure shown in Fig. 5.9, the local total output in a global region is divided into export supply and domestic supply according to a CET function. For all goods exported from a global region, a CET function is used to divide goods into those exported to China and those exported to other global regions. Eq. (64)~Eq. (69) describe the CET relation, where $WZ_{rr,i}$ is total output, $EE_{rr,i,oc}$ is exports to global regions other than China and $EC_{rr,i}$ is exports from region rr to China. $WE_{rr,i}$ is the composite export supply from region rr to China and other global regions. $WD_{rr,i}$ represents the domestic supply for region rr.

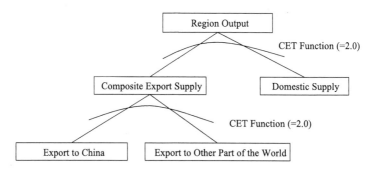

Fig. 5.9 Export Structure for World Region

$$WZ_{rr,i} = \alpha WDE_{rr,i} \left[\beta WD_{rr,i} WD_{rr,i}^{\frac{\sigma WDE+1}{\sigma WDE}} + \beta WE_{rr,i} WE_{rr,i}^{\frac{\sigma WDE+1}{\sigma WDE}} \right]^{\frac{\sigma WDE}{\sigma WDE+1}} \quad (64)$$

$$WE_{rr,i} = \left[\frac{\alpha WDE_{rr,i} \beta WE_{rr,i} (1 + \tau WZ_{rr,i}) PWZ_{rr,i}}{PWE_{rr,i}} \right]^{-\sigma WDE} \cdot \frac{WZ_{rr,i}}{\alpha WDE_{rr,i}} \quad (65)$$

$$WD_{rr,i} = \left[\frac{\alpha WDE_{rr,i}\beta WD_{rr,i}(1+\tau WZ_{rr,i})PWZ_{rr,i}}{PWD_{rr,i}}\right]^{-\sigma WDE} \cdot \frac{WZ_{rr,i}}{\alpha WDE_{rr,i}} \quad (66)$$

$$WE_{rr,i} = \alpha WE_{rr,i}\left[\sum_{oc \in OC}\beta EE_{rr,i,oc}EE_{rr,i,oc}^{\frac{\sigma WDE-1}{\sigma WDE}} + \beta EC_{rr,i}EC_{rr,i}^{\frac{\sigma WDE-1}{\sigma WDE}}\right]^{\frac{\sigma WDE}{\sigma WDE-1}} \quad (67)$$

$$EE_{rr,i,oc} = \left[\frac{\alpha WE_{rr,i}\beta EE_{rr,i,oc}PWE_{rr,i}}{(1-\tau EE_{rr,i,oc})PEE_{rr,i,oc}}\right]^{-\sigma WDE} \cdot \frac{WE_{rr,i}}{\alpha WE_{rr,i}} \quad (68)$$

$$EC_{rr,i} = \left[\frac{\alpha WE_{rr,i}\beta EC_{rr,i}PWE_{rr,i}}{(1-\tau EC_{rr,i})PEC_{rr,i}}\right]^{-\sigma WDE} \cdot \frac{WE_{rr,i}}{\alpha WE_{rr,i}} \quad (69)$$

Fig. 5.10 shows the import structure of the GTAP part of the model. Imports from China and other countries are first combined as the composite import for the local region under a CES function.

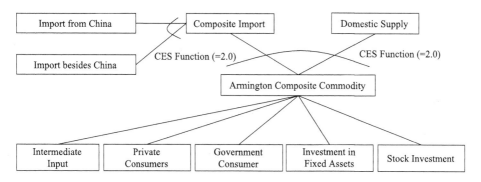

Fig. 5.10 Import Structure for World Region

$$WM_{rr,i} = \alpha WM_{rr,i}\left[\sum_{oc \in OC}\beta MM_{rr,i,oc}MM_{rr,i,oc}^{\frac{\sigma WM-1}{\sigma WM}} + \beta MC_{rr,i}MC_{rr,i}^{\frac{\sigma WM-1}{\sigma WM}}\right]^{\frac{\sigma WM}{\sigma WM-1}} \quad (70)$$

$$MM_{rr,i,oc} = \left[\frac{\alpha WM_{rr,i}\beta MM_{rr,i,oc}PWM_{rr,i}}{(1+\tau MM_{rr,i,oc})PMM_{rr,i,oc}}\right]^{-\sigma WM} \cdot \frac{WM_{rr,i}}{\alpha WM_{rr,i}} \quad (71)$$

$$MC_{rr,i} = \left[\frac{\alpha WM_{rr,i}\beta MC_{rr,i}PWM_{rr,i}}{(1+\tau MC_{rr,i})PMC_{rr,i}}\right]^{-\sigma WM} \cdot \frac{WM_{rr,i}}{\alpha WM_{rr,i}} \quad (72)$$

$$WQ_{rr,i} = \alpha WDM_{rr,i} \left[\beta WD_{rr,i} WD_{rr,i}^{\frac{\sigma WDM-1}{\sigma WDM}} + \beta WM_{rr,i} WM_{rr,i}^{\frac{\sigma WDM-1}{\sigma WDM}} \right]^{\frac{\sigma WDM}{\sigma WDM-1}} \tag{73}$$

$$WD_{rr,i} = \left[\frac{\alpha WDM_{rr,i} \beta WD_{rr,i} PWQ_{rr,i}}{PWD_{rr,i}} \right]^{-\sigma WDM} \cdot \frac{WQ_{rr,i}}{\alpha WDM_{rr,i}} \tag{74}$$

$$WM_{rr,i} = \left[\frac{\alpha WDM_{rr,i} \beta WM_{rr,i} PWQ_{rr,i}}{PWM_{rr,i}} \right]^{-\sigma WDM} \cdot \frac{WQ_{rr,i}}{\alpha WDM_{rr,i}} \tag{75}$$

$$PEE_{rr,i,oc} = \varepsilon_{rr} PWE_{rr,i} \tag{76}$$

$$PEC_{rr,i} = \varepsilon_{rr} PWE_{rr,i} \tag{77}$$

$$PMM_{rr,i,oc} = \varepsilon_{rr} PWM_{rr,i} \tag{78}$$

$$PMC_{rr,i} = \varepsilon_{rr} PWM_{rr,i} \tag{79}$$

Eq. (70)~Eq. (75) describe the import CES function, where $WQ_{rr,i}$ represents the Armington composite commodity. $MM_{rr,i,oc}$ denotes imports from other countries and $MC_{rr,i}$ denotes imports from other regions in China. $WM_{rr,i}$ represents total imports from other countries and other regions in China.

$$WQ_{rr,i} = \sum_{j \in J} WX_{rr,i,j} + WXH_{rr,i} + WXG_{rr,i} + WXI_{rr,i} \tag{80}$$

$$\sum_{i \in I} WLAB_{rr,i} = FWLAB_{rr} \tag{81}$$

$$\sum_{i \in I} WK_{rr,i} = FWK_{rr} \tag{82}$$

$$\sum_{i \in I} WLAN_{rr,i} = FWLAN_{rr} \tag{83}$$

$$\sum_{i \in I} WENE_{rr,i} = FWENE_{rr} \tag{84}$$

$$MM_{rr,i,oc} = EE_{oc,i,rr} \tag{85}$$

$$PWM_{rr,i} = PWE_{oc,i} \tag{86}$$

Eq. (80)~Eq. (86) represent the market-clearing conditions for the GTAP part of the model. Eq. (80) is the commodity market-clearing condition. Eq. (81)~Eq. (84) represent the balance of the labor, capital, land and natural resources markets. Eq. (85) and Eq. (86) describe the international trade balance for different global regions.

$$\sum_{r \in R} E_{r,i} = \sum_{rr \in RR} MC_{r,i} \tag{87}$$

$$\sum_{r \in R} M_{r,i} = \sum_{rr \in RR} EC_{r,i} \qquad (88)$$

$$PWE_i = PWM_{rr,i} \qquad (89)$$

$$PWM_i = PWE_{rr,i} \qquad (90)$$

Eq. (87)~Eq. (90) describe the conditions for linking MRCGE to GTAP. In Eq. (87), $\sum_{r \in R} E_{r,i}$ describes total exports from China to the rest of the world according to MRCGE and $\sum_{rr \in RR} MC_{r,i}$ describes Chinese imports from the rest of the world according to GTAP. Thus, Eq. (87) balances equation MRCGE exports and GTAP imports for China. Similarly, Eq. (88) balances MRCGE imports and GTAP exports for China. Eq. (89) and Eq. (90) balance the world market price between MRCGE export prices and GTAP import prices, and between MRCGE import prices and GTAP export prices, respectively. In these two equations, PWE_i is the MRCGE world market price for export commodities and $PWM_{rr,i}$ is the GTAP world market price for import commodities. PWM_i is the MRCGE world market price for import commodities and $PWE_{rr,i}$ is the GTAP world market price for export commodities.

5.2 Data and Scenario

5.2.1 Regional Division

(a) Regional divisions in China

For this research, 31 mainland provinces and municipalities in China were divided into eight regions based on geography and economic data. A detailed description can be found in Table 4.1 and Chapter 4.2.1.

(b) Global regions

The rest of world excluding mainland China was divided into nine different regions according to international trade relationships with China and ranking of CO_2 emissions. The regions are listed Table 5.1. OCN (Oceania), JPN (Japan), USA, EU27 (27-member EU) and RUS (Russian Federation) were treated as regions that include most of the industrialized nations of the world, viewed as the developed world in subsequent analysis.

Table 5.1 World's Regional Division

No.	Region	Country and Region Included
1	OCN	Australia, New Zealand, Rest of Oceania
2	JPN	Japan
3	GCA	Hong Kong, Taiwan
4	ROA	Rest of Asia
5	USA	United States of America
6	EU27	European Union 27 (Belgium, Bulgaria, Czechoslovakia, Denmark, Germany, Estonia, Greece, Spain, France, Ireland, Italy, Cyprus, Latvia, Lithuania, Luxembourg, Marta, Hungary, Netherlands, Austria, Poland, Portugal, Romania, Slovakia, Slovenia, Finland, Sweden, U.K.)
7	RUS	Russian Federation
8	IND	India
9	ROW	Rest of the world

Region IND represents India, an important developing nation. Hong Kong and Taiwan were independently treated as a region, denoted by GCA (greater Chinese area), because of their strong correlation with international trade with China. Region ROA represents the rest of Asia, in which the main economic power is the Association of Southeast Asian Nations (AESAN). ROW denotes the rest of the world. The regional divisions are shown in Picture 5.1.

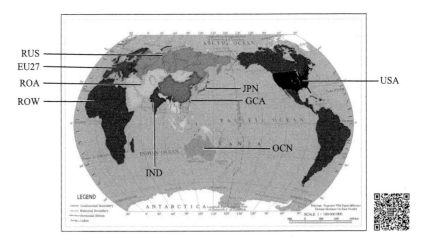

Picture 5.1   Regional Division of Worldwide

5.2.2   Industry Classification

Data for the MRCGE-GTAP model are based on the Chinese 2000 multi-regional IO matrix (2002) and the GTAP7 database (2002). The IO matrix includes eight regions and 30 commodity sectors. However, the present study investigates both the effects of energy tax policies and whether the policy adopted by a country could affect the world in the future.

For commodity data, we used database values and reclassified industry data sources. As shown in Table 5.2, the commodities were reclassified into 24 sectors.

Table 5.2 Reclassified Commodity Sectors

No.	New Classified Sectors	China Multi-region IO 30 sectors	GTAP7 sectors
1	Agriculture	Agriculture	pdr, wht, gro, v_f, osd, c_b, pfb, ocr, ctl, oap, rmk, wol, for, fsh
2	Coal Mining	Coal Mining	col
3	Oil and Gas Mining	Oil and Gas Mining	oil, gas
4	Other Mining	Metal Ore Mining, Non-Metal Ore Mining	omn
5	Food Manufacturing	Food Manufacturing and Tobacco Processing	cmt, omt, vol, mil, pcr, sgr, ofd, b_t
6	Textile	Textile	tex
7	Wearing Apparels	Wearing Apparel, Leather, Furs, Down and Related Products	wap, lea
8	Sawmills and Wood Products	Sawmills and Furniture	lum
9	Paper Products	Paper and Products, Printing and Recording Medium Reproduction	ppp
10	Petroleum Processing and Coking	Petroleum Processing and Coking	p_c
11	Chemical Industry	Chemical Industry	crp
12	Non-metallic Mineral Products	Non-Metallic Mineral Products	nmm
13	Metal Smelting and Pressing	Metal Smelting and Pressing	i_s
14	Metal Products	Metal Products	nfm, fmp
15	Machinery Industry	Machinery Industry	mvh
16	Transport Equipment	Transport Equipment	otn
17	Electrical Machinery and Equipment	Electrical, Machinery and Equipment	ome
18	Electronic and Communication Equipment	Electronic and Communication Equipment Manufacturing	ele
19	Other Manufacturing Industries	Measuring Instruments and Office Machinery, Machinery and Equipment Repair, Other Manufacturing Industries, Waste Disposal	omf

			Continued
No.	New Classified Sectors	China Multi-region IO 30 sectors	GTAP-7 sectors
20	Electricity, Water and Gas Supply	Electricity, Steam, and Hot Water Production and Supply, Gas Production and Supply, Tap Water Production and Supply	ely, gdt, wtr
21	Construction	Construction	cns
22	Transportation and Warehousing	Transportation and Warehousing	otp, wtp, atp, cmn
23	Commercial	wholesale and Retail Trade	trd
24	Services	Services	ofi, isr, obs, ros, osg, dwe

Unlike the data taken directly from the Chinese 2000 multi-regional IO matrix for the MRCGE model, IO data for Chinese regions were modified to provide a balance between GTAP and regional Chinese data.

The GTAP7 database provides global data. Although some of the IO tables are only updated to 1998 or 2002, the database uses bilateral international trade data for 2004. Therefore, it represents global economic flows for 2004. Data for the Chinese 2000 multi-regional IO matrix are based on the provincial IO table for 1997. To balance the GTAP7 and Chinese databases, data in the multi-regional IO table were extended to 2004 using the real GDP for China in 2004 and bilateral trade data for China in the GTAP7 database. Import and export data in the Chinese multi-regional IO table were extrapolated to 2004 using the ratio of international trade for one industry in one region to the overall trade for that industry. The process was repeated for each industry and each region. Other data in the multi-regional IO table were extrapolated using a multiple of real GDP growth normalized using actual GDP for China in 2004. In this process, no industry structure change was considered.

5.2.3 Scenarios

Using the MRCGE-GTAP model described above, we set up five different scenarios to evaluate the effects of a carbon tax levied in industrialized countries, China and the entire world. Scenario goals were the global CO_2 emission reduction target in the Kyoto Protocol and the commitment by major emitters made at the COP15 United Nations Climate Change Conference held in Copenhagen (Table 5.3). Although the first-period target of the Kyoto Protocol was not met, it is still the only international common goal for CO_2 reduction, so we used this compared to year 1990 levels multiplied by 5.2% (equal to comparison to world 2005 levels multiplied by 4.4%) as the global CO_2 reduction target in scenarios.

Table 5.3   Reduction Target for Developed Countries

Sector	Copenhagen Commitment		Kyoto Protocol Aim
	Compared to 1990 levels	Compared to 2005 levels	Compared to 1990 levels
Australia	4%~24% below	10%~29% below	8% above
Canada	23% above	17% below	6% above
European Union	20%~30% below	17%~28% below	7.7% below
Japan	25% below	29% below	6% below
New Zealand	10%~20% below	30%~38% below	remain
Norway	30%~40% below	24% below	1% above
Russia Federation	20%~25% below	23%~9% above	remain
Switzerland	20%~30% below	24%~34% below	8% below
USA	in the range of 4% below	in the range of 17% below	did not ratify

Data source: http://switchboard.nrdc.org/blogs/jschmidt/developed_country_emissions_targets.html

Besides the global target in the Kyoto Protocol, since emissions from the EU and the USA accounted for nearly one-third of the total global CO_2 emission in recent years,[①] and thus could represent the majority of emissions for all industrialized countries, we took the COP15 commitment for these two areas (compared with 2005, a reduction in CO_2 emission of ~17% for the USA and 14% for the EU) as the total reduction target for industrialized countries. The Kyoto Protocol target for total world reduction was set as the main goal for scenario simulation.

Table 5.4   Scenario Setup

Scenario	Content
S1	Industrialized countries execute same level carbon tax to reach the Kyoto Protocol Aim
S2	Industrialized countries execute same level carbon tax to main Industrialized countries Copenhagen commitment
S3	Industrialized countries and China execute same level carbon tax to reach the Kyoto Protocol Aim
S4	Industrialized countries and China execute same level carbon tax to main Industrialized countries Copenhagen commitment
S5	All nation's and the world execute same level carbon tax to reach the Kyoto Protocol Aim

---

① Global total CO_2 emission was 30313.248 million ton in 2009, with the USA accounting for 5424.53 million ton and the EU for 4307.285 million ton. More specific data are available on the US Energy Information Administration web page (http://www.eia.gov/cfapps/ipdbproject/IEDIndex3.cfm?tid=90&pid=44&aid=8).

In addition to the different objectives, the main difference between each scenario is the implementation of a carbon tax in different countries. In scenarios 1 and 2, only industrialized countries impose a carbon tax. In scenarios 3 and 4, industrialized countries and China impose a carbon tax. In scenario 5, there is global imposition of the same carbon tax to meet the CO_2 emission reduction target of the Kyoto Protocol.

5.3 Analysis Results

Changes in equivalent variation (EV), reduction rates for CO_2 emission and the rate of change in industrial output for Chinese regions were selected for analysis of economic and environmental effects for different Chinese and global regions. Indexes selected for global region analysis were the reduction rate for CO_2 emission and EV changes.

5.3.1 Changes in EV and CO_2 Emission Reductions in China

In scenarios 1 and 2, China does not impose a carbon tax and the EV for all regions in China remains almost unchanged, as shown in Fig. 5.11. In scenarios 3~5, China levies a carbon tax and each region shows a decrease in regional EV.

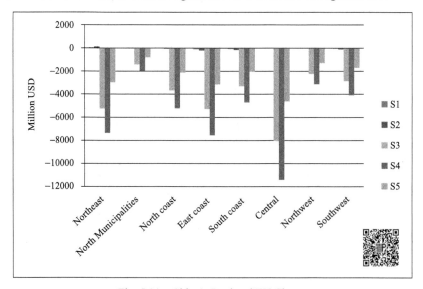

Fig. 5.11 China's Regional EV Change

For all regions, the largest EV decrease was observed in scenario 4, in which China and industrialized countries impose a carbon tax to reach the total reduction target for industrialized countries. The next largest EV decrease was observed in scenario 3, in which China and industrialized countries apply a carbon tax to achieve the world CO_2 emission reduction target. In Scenario 5, in which all countries impose the same carbon tax to comply with the Kyoto Protocol, Chinese regional EV losses were less than in scenarios 3 and 4.

In the three EV reduction scenarios, central China, where coal is mainly produced, had the largest EV decline, followed by the east coast, which has the most outsourced economy, and the northeast, which has the largest petroleum field in China. Among the eight regions, the EV decline was relatively low in northern municipalities; this might be because this region includes the cities of Beijing and Tianjin. Thus, population differences between this and other regions lead to a relatively small change in regional EV.

Since the EV change for China in scenarios 1 and 2 is positive, it is not surprising that CO_2 emission increases in China. Scenarios for Chinese regional reductions in CO_2 emission were selected for analysis. Fig. 5.12 shows the same trend as for regional EV changes. The south coast of China has the greatest CO_2 emission reduction in each scenario. In scenario 3, in which China and industrialized countries impose a carbon tax to reach the Kyoto Protocol target, CO_2 emission for the south coast of China decreased by 9%. In scenario 4, this index increased by a further 2%.

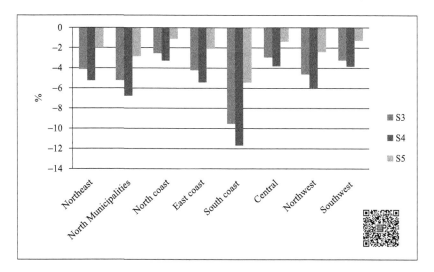

Fig. 5.12 China Regional CO_2 Emission Reduction Rate of Change

CO_2 emission reductions for other regions in China are lower than for the south coast, especially in central China, where the main coal production base is located and household utility in this region would suffer the most from an EV decrease. This situation can be explained as follows. A carbon tax levied in central China might have a negative effect on the regional economy, which would directly affect household utility. However, energy consumption in China is highly dependent on coal use (coal use has accounted for >70% of primary energy consumption for a decade), most of which comes from the domestic supply. Thus, as the major coal production area, central China could not stop investing in industries such as coal mining, so any carbon tax would be reinvest in coal production in the region. This reinvestment flow weakens the effect of the carbon tax and causes the above result.

5.3.2 China's Industry Changes

Fig. 5.13 provides an example of the economic effect of a carbon tax policy in China. It is clear that for every scenario in which China levies a carbon tax, a significant decrease in output is observed for energy-intensive industries such as petroleum processing and coking, the chemical industry and the energy supply industry, including coal mining, oil and gas mining, and electricity, water and gas supply; only low-energy industries such as services show an increase in output in these scenarios. However, in scenarios in which China does not levy a carbon tax, energy-intensive and energy supply industries in China show positive growth.

It should also be noted that the textile and clothing industries show a decrease in all scenarios, even those in which China does not levy a carbon tax. Since these are two of the most export-oriented industries in China, it is natural that when the world economy is affected by a carbon tax policy, these industries would also be affected, even if a carbon tax is not imposed in China.

5.3.3 Change of Household EV and CO_2 Reduction of World Regions

As shown in Fig. 5.14, for most scenarios in which a carbon tax is imposed by industrialized countries, regions in industrialized countries all show a significant EV decrease due to a carbon tax policy. Among industrialized regions, the USA and EU show the strongest EV decline, especially in scenarios 2 and 4, in which the COP15 commitment is the target. Developing and the least developed regions only show an

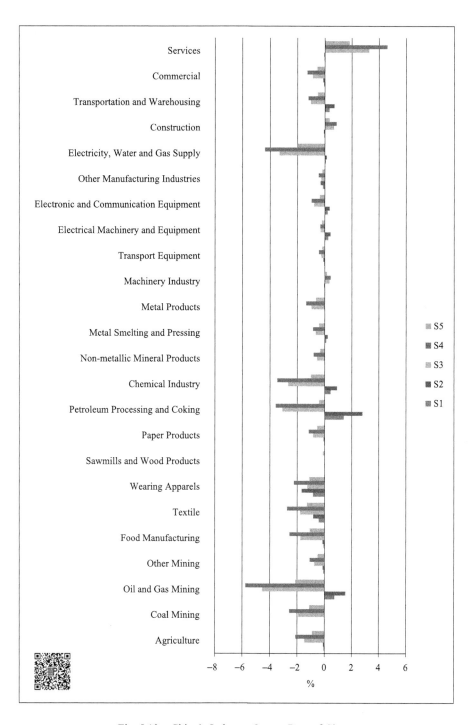

Fig. 5.13 China's Industry Output Rate of Change

EV reduction in scenario 5. The EV changes for all regions in the world explain the decrease in output for the textile and clothing industries in China: since industrialized regions are the most important importers of textile products from China, a decrease in household utility in these regions naturally leads to a decrease in the textile and clothing industries in China. This example provides strong evidence that when linked to international trade, no country can avoid the effects of such a global economic symbiosis.

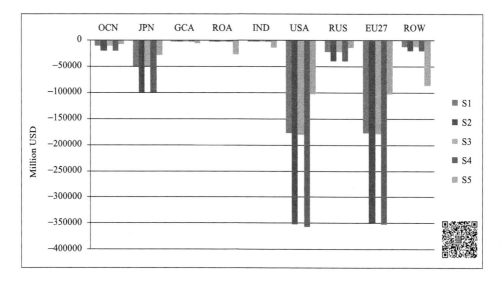

Fig. 5.14 World Region EV Change

Besides an EV decrease in household utility for different world regions, a carbon tax also leads to reduction in regional CO_2 emissions to meet the different scenario goals. The results in Fig. 5.15 show that when a region levies a carbon tax, its regional CO_2 emission decreases, while the opposite behavior might lead to an increase in regional CO_2 emission.

Fig. 5.15 also shows that if industrialized countries meet the COP15 commitment (scenarios 2 and 4), the overall reduction in global emission levels is much great than for the Kyoto Protocol target. Thus, a CO_2 reduction policy will be more effective for industrialized countries than for developing and less developed countries from a global perspective.

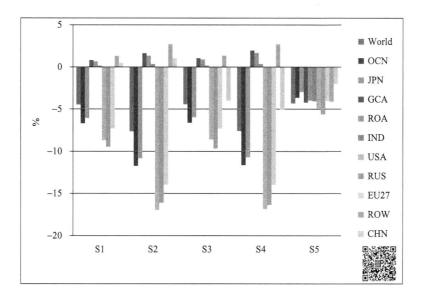

Fig. 5.15 World CO_2 Emission Reduction Rate under Different Scenarios

5.3.4 Industrial Output in World Regions

As in China, other regions experience changes under different scenarios for a carbon tax. Fig. 5.16~Fig. 5.24 show changes in industrial output for different world regions.

Fig. 5.16 shows the change in industrial output for OCN. It is evident that energy resource manufacturing and processing industries and the major industries of the region, such as agriculture and food manufacturing, show a decrease in output, whereas the output of other industries increases in each scenario. The largest decreases occur for scenarios 2 and 4, in which the major developed countries meet the COP15 commitment.

In scenarios 1 and 2, compared with a business as usual (BAU) scenario, OCN as a developed region shows a decrease of ~16% in petroleum processing and coking industries and of ~14% in oil and gas mining. This change in industrial output is caused by the carbon tax levied in the experimental scenarios and by international trade relations. However, it should be noted that for scenarios in which major industrialized countries meet either the Kyoto Protocol or the COP15 commitment, the change in industrial output would be smaller if China also levies a carbon tax. At the same time, EV for household utility in OCN hardly changed. It can be concluded that OCN would benefit by imposition of a carbon tax in China.

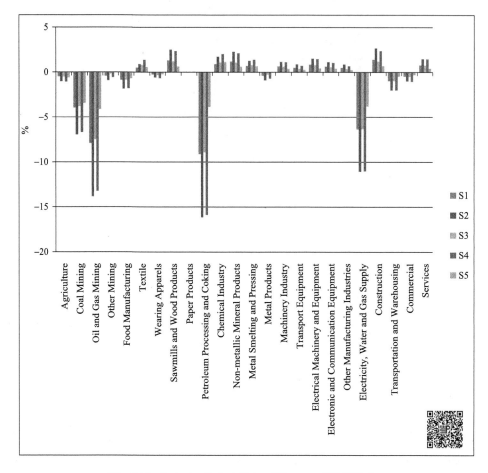

Fig. 5.16　Industry Output Rate of Change of "OCN" Area

Fig. 5.17 shows changes in industrial output for JPN under the five different scenarios. Since Japan was identified as a developed country but without strict GHG control for the scenario set-up, it shows almost the same behavior for changes in industrial output rate as OCN. Decreases are observed for energy resource manufacturing and processing industries and the major industries in Japan (machinery and electric machinery and equipment), while increases are observed for most other industries in all scenarios. As for OCN, Japan would also benefit if China were to levy a carbon tax along with the developed countries. As a result, changes in industrial output for Japanese energy resource manufacturing and processing industries and its major industries would be smaller, while its EV for household utility would hardly change.

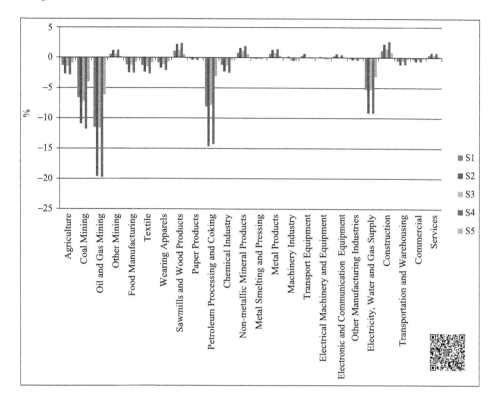

Fig. 5.17 Industry Output Rate of Change of "JPN" Area

Fig. 5.18 shows changes in industrial output for GCA. Decreases in output for energy resource manufacturing and processing industries, apart from coal mining, are only observed in scenario 5, in which the whole world has the same carbon tax. Industries such as textile and clothing, which are the main export industry in China, show the same trends as for these industries in China.

As previously mentioned, the GCA area has high correlation in trade with China. Moreover, many Chinese commodities are re-exported from the GCA area, especially Hong Kong. It is not surprising that industrial output for a sector decreases in the GCA area when the output for the same sector in China decreases. It should also be noted that energy-intensive industries in GCA, such as petroleum processing and coking and the chemical industry, all increase in scenarios 1~4. It can be concluded that energy-intensive industries in GCA might benefit when other major countries apply a carbon tax but GCA does not.

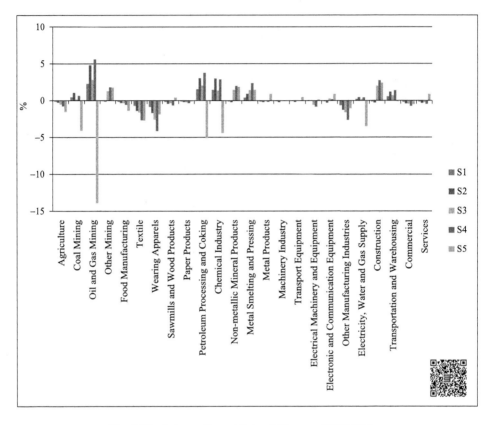

Fig. 5.18 Industry Output Rate of Change of "GCA" Area

Fig. 5.19 shows changes in industrial output for IND. As for the three cases above, India would experience a decrease in output for energy resource manufacturing and processing industries if it were to levy a carbon tax.

One interesting observation for India is that most of its secondary and tertiary industries that are not energy-intensive show an increasing trend compared with the BAU scenario. In fact, such universal growth in output is also observed for various industries in ROA and ROW. This is not surprising, since these three regions include less developed countries. Such universal growth represents a benefit and an opportunity for development in less developed countries arising from a global GHG control policy.

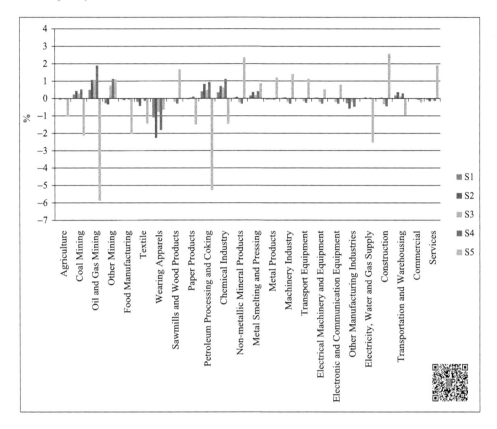

Fig. 5.19 Industry Output Rate of Change of "IND" Area

Fig. 5.20 shows changes in industrial output for ROA. The major economic activity in this area is oil production in countries in Central Asia, AEASN and other developing or less developed countries of Asia. Energy resource manufacturing and processing industries in this region show a decrease in output when ROA does not levy a carbon tax.

Moreover, since ROA includes less developed countries, the region benefits from development opportunities arising from a global GHG control policy. The chance was proved by the increase of the majority of non-energy resource-intensive industries while the worldwide carbon tax was arranged.

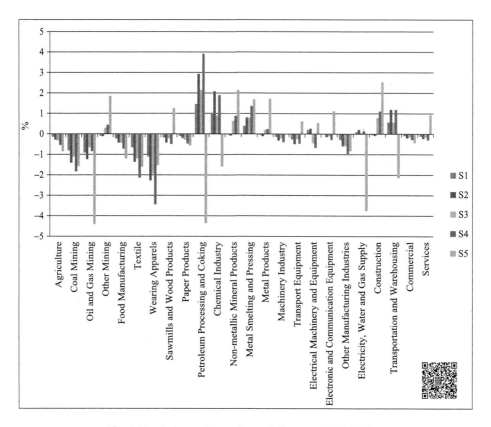

Fig. 5.20 Industry Output Rate of Change of "ROA" Area

Fig. 5.21 shows changes in industrial output for RUS. The rate of change for many industries is quite significant compared with other regions. In fact, changes in industrial output, especially for energy resource manufacturing and processing industries, was only smaller than changes in the USA and EU27, the two major regions for each scenario. Moreover, for the electricity, water and gas supply sector, RUS had the largest reduction among all regions. This might be attributable to the economic structure of the Russian Federation.

Natural gas is one of the main exports from Russia, the majority of which is exported to European countries. In fact, a quarter of the annual natural gas demand in European countries is imported from Russia.[①] This trade relation means that the

---

① As described in the news report "Russia's gas pipeline to Europe", Xinhua Net 2009. url:http://news.xinhuanet.com/world/2009-01/08/content_10623580.htm (in Chinese).

Russian economy is highly correlated to economic development in Europe. When the EU applies a carbon limit policy as in the simulation scenarios, a decrease in EU energy resource use would naturally lead to a reduction in productivity for the main supplier of one of its energy resources. As a result, a significant decrease is observed for energy resource manufacturing and processing industries in Russia.

According to the results, it can be concluded that energy-intensive industries in Russia would experience a negative effect on their economic development if a carbon tax is applied, even if the tax is not levied in Russia.

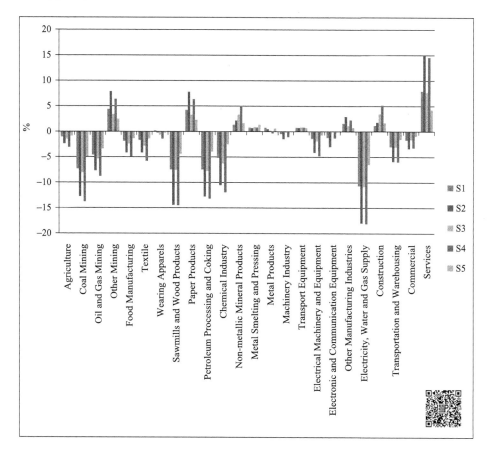

Fig. 5.21 Industry Output Rate of Change of "RUS" Area

Fig. 5.22 shows changes in industrial output for the USA. The changes are similar to those for OCN and Japan: decreases are evident for energy resource manufacturing and processing industries and energy-intensive industries, while other industries show only a small rate of change (<5%).

Since the USA is a key control area in each scenario, huge reductions were observed for industries related to energy resources.

As observed for other developed countries, the USA benefits if China levies a carbon tax. However, levy of a carbon tax in China does not change output for the electricity, water and gas supply sector in the USA, whereas this sector is affected in OCN and Japan.

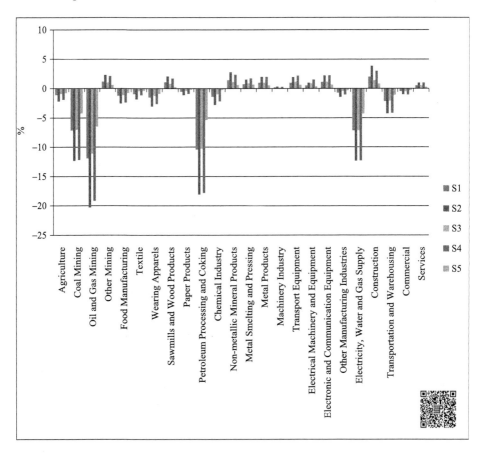

Fig. 5.22   Industry Output Rate of Change of "USA" Area

Fig. 5.23 shows changes in industrial output for EU27, which are similar to those for the USA. The only difference is in the ratio of the number, especially for sectors that are not energy-intensive. Since both the EU and the USA are treated as key policy control areas, this trend is as expected.

The EV change in household utility differs for the EU and the USA. Although the change in energy resource manufacturing and processing industries was slightly greater for the EU than for the USA, it is possible that this difference arises from differences in industry structure between the two regions.

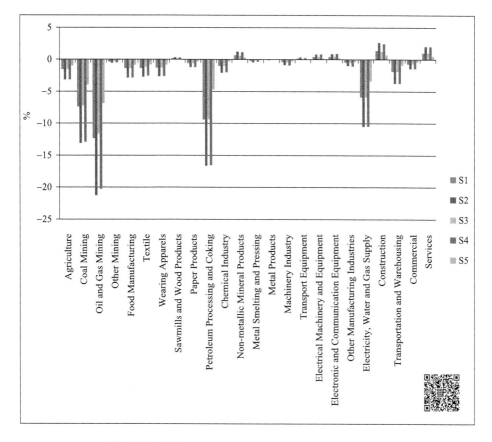

Fig. 5.23 Industry Output Rate of Change of "EU27" Area

Finally, Fig. 5.24 shows changes in industrial output for ROW. Most industries show an increase in for scenarios 1~4.

The most significant changes are observed for the petroleum processing and coking industry in scenarios 2 and 4. This can be explained as follows. Although energy-producing industries such as petroleum processing and coking are limited by a carbon tax policy in some regions of the world, petroleum is still in high demand as an energy resource. The supply gap caused by policy limitation could only be filled by regions without a GHG control policy.

Under this market regulation, the petroleum processing and coking industry in ROW gains an opportunity for own development. The market mechanism also benefits energy-intensive industries such as the chemical industry in ROW. Thus, economic promotion of less developed and developing regions is driven by GHG reduction policies, as mentioned above.

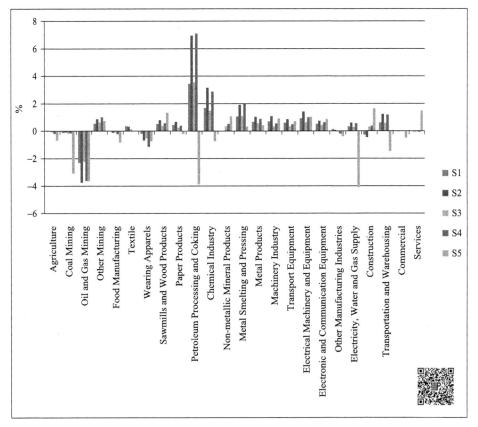

Fig. 5.24 Industry Output Rate of Change of "ROW" Area

5.4 Conclusion

A MRCGE-GTAP model was applied to analyze the effect of a carbon tax in different world regions on China and the global economy. Five conclusions can be drawn from the simulation results.

First, the results from each scenario show that application of a carbon tax could effectively reduce CO_2 emissions worldwide. Furthermore, if the main industrialized countries applied the same level of tax, the Kyoto Protocol target is achievable. Moreover, if these countries could meet their COP15 commitment, the Kyoto Protocol target could be achieved to an even greater extent. However, it should be noted that total CO_2 was endogenous for energy-producing and energy-intensive sectors in the MRCGE-GTAP model, and thus reductions observed for the simulations could be considered as reductions due to industry activities.

Second, a carbon tax might have negative effects on major developed countries, especially for Japan, the USA and the EU. The simulation results show that these regions would suffer the greatest household utility losses while making the greatest contributes to CO_2 emission reductions.

Combining the above conclusions, two issues are clear: on one hand, industrialized countries are the main agents in achieving global GHG reductions; on the other hand, the more successful that reduction is, the greater will be the negative effect it has in citizens in industrialized countries. Thus, an appropriate reduction policy should balance the pros and cons for GHG reductions in industrialized countries.

Third, a worldwide carbon tax that is not levied in China may benefit Chinese households by increasing their consumption utility. However, this policy might have a negative effect on the main export industries in China. As observed for scenarios 1 and 2, China would benefit from a carbon tax levied in other countries, but since such a tax would affect the major nations in the global economy, countries that benefit from globalization such as China would experience a decrease in their main international trade industries, as shown in Fig. 5.13. This result indicates that in the current global economy, no country can escape facing a global issue as climate change: even if a country does not participate in global attempts to reduce CO_2 emissions, its economy will still be affected.

Fourth, the simulations show that levy of a carbon tax in China will cause a decrease in household consumption utility in different regions of the country. Household utility in regions that produce and outsource energy resources would suffer the most. Thus, when a carbon tax is levied nationwide in China, regional differences should be considered as if the levy was worldwide.

Finally, output analysis for a range of industries showed that less developed countries will benefit from a carbon tax, regardless of whether it is levied in developed countries only or worldwide. At the same time, the economy of resource-oriented countries with international trade may be damaged. The results show that most countries will gain from a carbon tax in developed countries, whether the policy has the Kyoto Protocol or COP15 commitment as a target. Although developed countries, especially the USA and EU, suffer the most in each scenario, they will experience less of a negative impact if China levies a carbon tax.

Concluding Remarks

6.1 Simulation Model Construction

In this study, a one-country multi-regional CGE model was constructed based on the 2000 multi-regional IO matrix for China. The model was used to evaluate the effect of an energy resource policy in China. Using the GTAP7 database, the multi-regional model was expanded to a CGE model combined with a GTAP model. This new MRCGE-GTAP model facilitated an analysis of the effect of a carbon tax on different regions in China and throughout the world.

This research is the first to evaluate the potential effects of a carbon tax in China on environmental and regional economic development in both the country and worldwide. This represents the most innovative contribution of the study. However, the CGE models used were static models. In CGE studies, a dynamic CGE model undoubtedly provides more convincing and reliable results than a static model. Therefore, future research will expand the model to a dynamic model for policy analysis.

Another important issue for future research is to improve the current MRCGE-GTAP model. Considering inter-regional trade as an example, transport costs were not considered for Chinese or global regions in the current model, but these costs should be carefully considered for real economic situations. Therefore, transports costs for regional trade should be specified in a future study.

The MRCGE-GTAP model constructed for this study is highly reliant on different types of data. World data for 2004 from the GTAP7 database were used and the multi-regional IO matrix was even older; raw data for this matrix are Chinese 1997 province-level IO tables. These data are too old for a policy simulation tool such as CGE for policy analysis requiring strong timeliness. Thus, newer data will be required in future studies to improve the current MRCGE-GTAP model.

The scenarios may also need to be reconsidered as a future improvement for MRCGE-GTAP research. The two models used here are static CGE models, so issues such as dynamic economic growth and population growth are not taken into consideration. For energy resource usage, CO_2 emissions or even the Chinese COP15 commitment, the impact of such dynamic factors cannot be ignored. Thus, in future analysis with a dynamic model, the analysis scenarios should be reconsidered and new key conditions should be identified.

6.2 Analysis for China

A one-country, multi-regional CGE model and a multi-national, multi-regional CGE model were established to assess the impact of a carbon tax on resources in China and the validity and fairness of worldwide carbon tax, respectively. In the analysis of an energy resource tax in China (Chapter 4), Chinese CO_2 emissions in the base year were used directly as an exogenous variable. However, in the global multi-regional analysis of Chapter 5, because of sufficient data support from the GTAP database, CO_2 emissions for each economy could be directly calculated within the model. A direct manifestation of this difference is that for the one-country multi-regional CGE analysis, a 45% per unit GDP reduction in CO_2 emissions can be realized using CO_2 emissions in the base year as the standard. By contrast, in the MRCGE-GTAP analysis, even if the Kyoto Protocol targets were achieved worldwide or China and developed countries jointly implemented the COP15 commitments, the CO_2 emission reduction in China was ~5%, indicating a slight but positive effect of a carbon tax in China. The reason is that the potential 45% decrease in China is actually a slowly increasing target rather than a target calculated according to the current GDP growth rate in China. (If we consider that GDP growth in China remains >8% for 2005~2020, which is lower than the average for 2005~2011, an actual CO_2 reduction of 45% per unit GDP implies a total CO_2 increase by a factor of 1.74 compared with 2005.) Thus, the effect of a carbon tax on emission reductions is positive.

Besides the total reduction effect, this study also explored the issue of regional fairness under a carbon tax in China for both models. Results from the two models show that a carbon tax will have differential regional effects on household utility losses since there are large differences in development among Chinese regions. Regions with the greatest development and those that supply energy resources would suffer the largest household utility losses. In terms of economic development, regions that supply energy resources are usually less developed. Therefore, household utility losses for these regions would lead much worse living conditions compared with rich areas in China. Thus, for real policy application, the problem of policy fairness should be considered. For example, all new taxes levied were used to fund regional government consumption and investment in both models of the study. This activity was based on real regional government activity in China. The simulation results show that less economically developed areas would suffer more from a decrease in household utility. Therefore, if an energy resource tax is extended nationwide in China, a different tax level for different regions should be considered. For more developed regions, the new levy could be used to fund government consumption and investment. For less developed areas, the new levy should rather be considered as a transfer payment to the household sector to subsidize household utility. The simulation results from the two models provide evidence to support such a policy of different levels of tax by region.

The evidence that the carbon tax should differ by region in China is supported by the regional CO_2 reduction rate. In two different analyses, the rate of CO_2 reduction was different for every region in China. In MRCGE analysis, since the exogenous shock directly affects energy-producing industries, reductions in CO_2 emission were treated as a measure of reductions in output for these industries (the lower the industrial output, the less CO_2 is produced by that source). MRCGE results for petroleum and natural gas mining indicate that the most significant reduction in output (CO_2 reduction) occurs for the south coast of China. For coal mining, CO_2 reduction was better for central and northwest China that for other areas. The MRCGE-GTAP results show that the best direct CO_2 emission reduction was observed for the south coast. The simulation results suggest that whether China needs to apply the same level of tax in all regions is still in doubt. Analysis results for the two models indicate that a carbon tax should be levied in the most effective regions. For less effective regions, levying of such a new tax should be carefully considered.

Combining the two points mentioned above, tax rates should differ by region in China according to regional differences in household utility losses and the effects of CO_2 emission reduction for each region.

6.3 Analysis for the World

According to the results presented in Chapter 5, a worldwide carbon tax, especially levied in industrialized nations, could effectively reduce global carbon emissions and meet the Kyoto Protocol goal. Industrialized countries should take more responsibility for successful global reductions in CO_2 emissions.

The suggestion of different carbon tax rates for different regions in China should also be considered for different regions in the world. Although similar terms are included in the Kyoto Protocol, careful consideration is still required when executing policies to reduce climate change.

When considering the effect on world regions besides China, analysis revealed that even if a carbon tax is levied in only a few countries, the economic effects will be worldwide. As shown in Fig. 5.13, for scenarios 1 and 2, in which China does levy a carbon tax, industrialized countries experience a decrease in household utility and consequently the main export industry in China (clothing and textiles) suffers a decrease in output. Fig. 5.14 shows that although most world regions do not levy a carbon tax in scenarios 1~4, ROW, which includes most of the countries in the world, experiences a decrease in household utility. It can be concluded that no country can escape reductions in GHG emissions.

6.4 Conclusion

The situation regarding reductions in CO_2 emissions is not optimistic. The right of every country to develop and the need to reduce CO_2 emission seem to pose an irreconcilable conflict. However, there is only one earth and this needs to be a fundamental consideration for technology improvements and economic growth. If errors are made, there will be no second chance.

The study results show that effective actions should be immediately implemented to save the planet. There is truly not much time left.

Reference

Abler D G, Rodriguez A G, Shortle J S. 1999. Parameter uncertainty in CGE modeling of the environmental impacts of economic policies. Environmental and Resource Economics, 14(2): 75~94.

Adams P D, Parmenter B R, Horridge J M. 2000. Analysis of green house policy using MMRF-GREEN. Melbourne: Centre of Policy Studies, Monash University.

Armington P S. 1969. A theory of demand for products distinguished by place of production. International Monetary Fund Staff Paper, 16: 159~176.

Ballard C L. Fullerton D. Shoven J B, et al. 1985. A general equilibrium model for tax policy evaluation. Chicago: The University of Chicago Press.

Ban K. 2007. Multi-regional dynamic computable general equilibrium model of Japanese economies: forward looking multi-regional analysis. RIETI Discussion Paper Series, 2007-J-043.

Barker T, Baylis S, Madsen P. 1993. A UK carbon/energy tax – the macroeconomic effects. Energy Policy, 296~308.

Bergman L. 1990. Energy and environmental constraints on growth: a CGE modeling approach. Journal of Policy Modeling, 12(4): 671~691.

Bergman L. 1991. General equilibrium effects of environmental policy: a CGE-modeling approach. Environmental and Resource Economics, 1: 43~61.

Bernard A, Vielle M, Viguier L. 2003. Carbon tax and international emissions trading: a Swiss perspective. NCCR-WP4 Working Paper: 21.

Böhringer C, Rutherford T F. 1997. Carbon taxes with exemptions in an open economy: a general equilibrium analysis of the German tax initiative. Journal of Environmental Economics and Management, 32: 189~203.

Borges A M, Goulder L H. 1997. Decomposing the impact of higher energy prices on long-term growth. Cambridge: Cambridge University Press.

BP. 2011. Statistical review of world energy 2010. BP Global.

Burniaux J M, Truong T P. 2002. GTAP-E: an energy-environmental version of the GTAP model. GTAP Technical Paper, 16.

Conrad K, Henseler-Unger I. 1986. Applied general equilibrium modeling for long-term energy policy in Germany. Journal of Policy Modeling, 8(4): 531~549.

Conrad K, Schröder M. 1993. Choosing environmental policy instruments using general equilibrium models. Journal of Policy Modeling, 13(5): 521~543.

Department of Energy Statistics, National Bureau of Statistics. 2009. China energy statistical yearbook 2008. Beijing: China Statistics Press.

Dessus S, O'Connor D. 2003. Climate policy without tears: CGE-based ancillary benefits estimates for Chile. Environmental and Resource Economics, 25: 287~317.

Dixon P B, Parmenter B R, Sutton J, et al. 1982. ORANI: a multi-sectoral model of the Australian economy. Southern Economic Journal, 50(1): 126~126.

Gao P F, Chen W Y, He J K. 2004. Marginal carbon abatement cost in China. Journal of Tsinghua University, 44(9): 1192~1195.

Gopalakrishnan B N, Walmsleyeds T L. 2008. Global trade, assistance, and production: the GTAP7 data base. Center for Global Trade Analysis, Department of Agricultural Economics, Purdue University.

Gottinger H W. 1998. Greenhouse gas economic and computable general equilibrium. Journal of Policy Modeling, 20(5): 537~580.

Hatano T, Okuda T. 2006. Water resource allocation in the Yellow River basin, China applying a CGE model. Sendai: Intermediate Input-Output Conference.

Hayashiyama Y, Abe M, Muto S. 2012. Evaluation of GHG discharge reduction policy: by 47 prefectures multi-regional CGE. Journal of Applied Regional Science.

He J H, Shen K T, Xu S L. 2002. Carbon taxation and CGE model of carbon dioxide mitigation. Quantitative and Technical Economics, 19(10): 39~47.

Hertel T W. 1997. Global trade analysis: modeling and application. Cambridge: Cambridge University Press.

Horridge M, Wittwer G. 2008. SinoTERM, a multi-regional CGE model of China. China Economic Review, 19: 628~634.

Horridge M. 2000. ORANI-G: A general equilibrium model of the Australian economy. Preliminary Working Paper No. OP-93, Centre of Policy Studies and IMPACT Project, Monash University.

Hosoe N, Gasawa K, Hashimoto H. 2004. Textbook of computable general equilibrium modeling. Tokyo: University of Tokyo Press.

Houghton J T, Filho G M, Griggs L, et al. 1997. Stabilization of atmospheric greenhouse gases: physical, biological and socio-economic implications. Ipcc Technical Paper III Ipcc Secretariat.

Ianchovichina E, McDougall R. 2000. Structure of dynamic GTAP. GTAP Technical Paper, 18.

Ichimura S, W H J. 2004. Interregional Input-Output Analysis of the Chinese Economy. Sobunsha Press.

Institute of Developing Economies, Japan External Trade Organization. 2003. Multi-Regional Input-Output Model for China 2000. Chiba: Institute of Developing Economies, Japan External Trade Organization.

IPCC. 2006. 2006 IPCC guidelines for national greenhouse gas inventories, introductions. IPCC.

Jensen J, Toole R O, Matthews A. 2003. Controlling greenhouse gas emissions from agricultural sector in Ireland: a CGE modeling approach. Netherlands: Conference paper for 6th Annual

Conference on Global Economic Analysis.

Jorgenson D W, Wilcoxen P J. 1993. Reducing US carbon emissions: an econometric general equilibrium assessment. Resource and Energy Economics, 15: 7~25.

Kamat R, Rose A, Abler D. 1999. The impact of a carbon tax on the Susquehanna River Basin economy. Energy Economics, 21: 363~384.

Lau M I, Pahlke A, Rutherford T F. 2002. Approximating infinite-horizon models in a complementarily format: a primer in dynamic general equilibrium analysis. Journal of Economic Dynamic & Control, 26: 577~609.

Lee H L. 2002. An emissions data base for integrated assessment of climate change policy using GTAP. Purdue University, GTAP Working Paper (Draft).

Li S T, He J W. 2005. A three-regional computable general equilibrium (CGE) model for China. Beijing: 15th International Input-Output Conference.

Ministry of Environmental Protection of the People's Republic of China. 2010. China environment bulletin 2009. Ministry of Environmental Protection of the People's Republic of China.

Nestor D V, Pasurka C A. 1995. CGE model of pollution abatement processes for assessing the economic effects of environmental policy. Economic Modeling, 12: 53~59.

Pang J, Fu S. 2008. General equilibrium analysis in environmental economics: model, method and application. Beijing: Economic Science Press.

Pang J, Zou J, Fu S. 2008. Analysis on economic impact of China's fuel tax by using CGE model. Inquiry into Economic Issues, 11: 69~73.

Peter M W, Horridge M, Meagher G A, et al. 1996. The theoretical structure of MONASH-MRF. Centre of Policy Studies/IMPACT Centre Working Papers, 85.

Pezzey J, Lambie R. 2001. CGE Model for evaluating domestic greenhouse policies in Australia: a comparative analysis. Consultancy Report, Oridyctuvuty Commission.

Philip D J, Horridge J M, Parmenter B R. 2000. MMRF-GREEN: a dynamic, multi-sectorial, multi-regional model of Australia. Centre of Policy Studies/IMPACT Centre Working Papers, 94.

Pu Z N. 2011. A study on China's energy tax policy using the SCGE model. Discussion Paper, Tohoku University.

Pu Z N, Hayashiyama Y. 2012. Energy resource tax effects on China's regional economy by SCGE model. Environmental Economics, 3(1): 41~52.

Roland-Holst D, van der Mensbrugghe D. 2009. General equilibrium technique for policy modeling. Li S T, Duan Z G, Hu F, Translate. Beijing: Tsinghua University Press, 121-137(in Chinese).

Rosenthal R E. 2007. GAMS – a user guide. GAMS Development Corporation.

Rutherford T F, Paltsev S V. 2000. GTAP in GAMS and GTAP-EG: global datasets for economic research and illustrative models. Department of Economics, University Colorado, Working Paper: 1~64.

Sato T, Hino S. 2005. A spatial CGE analysis of road pricing in the Tokyo Metropolitan Area.

Journal of the Eastern Asia Society for Transportation Studies, 6: 608~623.

Scrimgeour F, Oxley L, Fatai K. 2005. Reducing carbon emissions? The relative effectiveness of different types of environmental tax: the case of New Zealand. Environmental Modeling & Software, 20: 1439~1448.

Shoven J B, Whalley J. 1984. Applied general equilibrium models of taxation and international trade. Journal of Economic Literature, 22: 1007~1051.

Shoven J B, Whalley J. 1992. Applied general equilibrium. Cambridge: Cambridge University Press.

Takashi K, Ryuichi S, Tomoki I, Ma L Q. 2005. Impacts of economic and transport policy partnership on international maritime container transportation using computable general equilibrium model. Technical Note of National Institute of Land and Infrastructure Management, 258.

Taylor L, Black S L. 1974. Practical general equilibrium estimation of resources pulls under trade liberalization. Journal of International Economics, 4(1): 37~58.

Thissen M. 2004. REAM 2. 0: a regional applied general equilibrium model for Netherlands. TNO Working Paper.

van der Mensbrugghe D. 2005. Linkage technical reference document: version 6. 0. Development Prospects (DECPG), the World Bank.

Wang C, Chen J N. 2006. Parameter uncertainty in CGE modeling for the macroeconomic impact of carbon reduction in China. Journal of Tsinghua University, 11(5): 617~624.

Wang C, Chen J N, Zou J. 2005. Impact assessment of CO2 mitigation on China economy based on a CGE model. Journal of Tsinghua University, 45(12): 1621~1624.

Wei W X. 2009. An analysis of China's energy and environmental policies based on CGE model. Statistical Research: 3~13.

William D N, Yang Z L. 1996. A regional dynamic general- equilibrium model of alternative climate- change strategies. The American Economic Review, 86 (4): 741~765.

Wissema W, Dellink R. 2007. AGE analysis of the impact of a carbon energy tax on the Irish economy. Ecological Economics, 61: 671~683.

Xia C W, Liu Y W. 2010. A dynamic CGE research on influence of China's energy saving emission reduction in the fuel tax levied. On Economic Problems, 2: 64~69.

Xie J, Saltzman S. 2000. Environmental policy analysis: an environmental computable general-equilibrium approach for developing countries. Journal of Policy Modeling, 22(4): 453~489.

Yang L, Mao X Q, Liu Q, et al. 2009. Impact assessment for energy taxation policy based on a computable general equilibrium model. China Population Resource and Environment, 19(2): 24~29.

Zhang X G. 2009. General equilibrium models: theory and application. Beijing: China Renmin University Press.

Zhao Y, Wang J F. 2008. CGE model and its applications in economic analysis. Beijing: China Economic Publishing House.

Appendix

For the last part of this book, I am providing all the main codes for three models that have been used for this research. Hence, here's this long borning appendix. Honestly speaking, consider those hard years for coding and debugging, these codes could be seen as the "real" research outcome for my three year Ph.D study life in Sendai, Japan.

Since the coding work during my study period followed all the "101" lessons about CGE coding, the following code can follow the basic logic from simple to complex, from static to dynamic in CGE model programming, which also can be used as a reference for new researchers who is willing to program their own CGE model program or have relevant research work.

Of course, these non-concise codes must look funny in the eyes of the experts of CGE modelling, so I really do hope to receive advises for my future work in this research area. As we always mentioned in any research articles: all errors remain to me. But there is one thing I do need to point out: for the market clearing module for global economy model, there are a lot of redundant computing language expressions. These are not because the author do not know that some simple loop statements can be used, but actually lead by excessive ambition that hope such model encompass all the circumstances, which directly caused data matching error occurred. Hence, the author can only list market clean conditions one by one to find the problem.

While finishing this appendix, those scenes like squatting on a chair to debug codes, solving model linkage structure with my colleague Mr. Abe Masahiro, arguing with Prof Hayshiyama for better computational device once again appeared in front of me. Although my supervisor Prof Hayashiyama has passed by more than three years, those days while studying with him will always be remembered in my heart. Since this is an academic book, I won't leave any postscript. But I hope the readers could allow me to leave these few words in this appendix, in memory of my professor, Yasuhisa Hayashiyama.

Model 1: One National, Multi-Region, Static Model

Codes based on GAMS 23.0 Software for Models
China Multi-Regional Model BAU Solution:
$TITLE CHINA MULTI REGION CGE MODEL

*DEFINITION OF SETS FOR SUFFIX

```
set
u/noea,nomu,noco,eaco,soco,cent,nowe,sowe,

agr,cm,ogm,om,fm,tex,wa,swp,pp,ppc,ci,nmmp,msp,mp,mi,te,em
e,ece,omi,ewgs,cons,tw,com,ser,

l,k,pt,dt,hoh,gov,inv,ex,im,ops/

r(u)/noea,nomu,noco,eaco,soco,cent,nowe,sowe/

i(u)/
agr,cm,ogm,om,fm,tex,wa,swp,pp,ppc,ci,nmmp,msp,mp,mi,te,em
e,ece,omi,ewgs,cons,tw,com,ser/



alias (U,V),(R,S,W),(I,J,H);
```

*LOADING DATA
```
table i_m(r,i,s,j)
$INCLUDE IntermediateInput.INC
```

Appendix 113

;

```
table f_d(r,i,s,v)
$INCLUDE FinalDemand.INC
;

table v_a(u,s,j)
$INCLUDE ValueAdded.INC
;

TABLE EXT(R,I,V)
$INCLUDE Trade.INC
;
```

*LOADING THE INITIAL VALUE

```
parameter
y0(s,j)            COMPOSITE FACTOR
l0(s,j)            LABOR
k0(s,j)            CAPITAL
xx0(r,i,s,j)       INTERMEDIATE INPUT
x0(i,s,j)
z0(s,j)            OUTPUT
xh0(r,i,s)
xg0(r,i,s)         GOVERNMENT CONSUMPTION
xi0(r,i,s)
tz0(s,j)           PRODUCTION TAX
epsilon0           EXCHANGE RATE
e0(r,i)            EXPORT
m0(r,i)            IMPORT
d0(r,i)            DOMESTIC
q0(r,i)            ARMINGTON
tauz(s,j)          PRODUCTION TAX RATE
taud
```

```
sf(r)              FOREIGN SAVING IN US DOLLAR
fl(s)              FACTOR ENDOWMENT OF LABOR
fk(s)              FACTOR ENDOWMENT OF CAPITAL
tr0(s)             TRANSFER
ex0(r)
im0(s)
sp0(s)
sg0(s)
td0(s)
ps(s)
pg(s)
ss(i,j)
;

l0(s,j)     = V_A("L",s,j);
k0(s,j)     = V_A("K",s,j);
tz0(s,j)    = V_A("PT",s,j);
y0(s,j)     = l0(s,j)+k0(s,j);
xx0(r,i,s,j) = I_M(r,i,s,j);
x0(i,s,j)   = sum(r,xx0(r,i,s,j));
z0(s,j)     = y0(s,j)+sum(i,x0(i,s,j));
xh0(r,i,s)  = F_D(r,i,s,"hoh");
xg0(r,i,s)  = F_D(r,i,s,"gov");
xi0(r,i,s)  = F_D(r,i,s,"inv");
fl(s)       = sum(j,l0(s,j));
fk(s)       = sum(j,k0(s,j));
tauz(s,j)$(z0(s,j))     = tz0(s,j)/z0(s,j);
*tauz(s,j)$(not z0(s,j))  = 0;
Taud = 0.15;
td0(s)      = taud*[fl(s)+fk(s)];
epsilon0 = 1;
e0(r,i)     = ext(r,i,"ex");
m0(r,i)     = (-1)*EXT(r,i,"im");
d0(r,i)     = (1+tauz(r,i))*z0(r,i)-e0(r,i);
```

```
q0(r,i)  = sum(s,xh0(r,i,s))+sum(s,xg0(r,i,s))+sum(s,xi0 (r,i,s))
+sum((s,j),xx0(r,i,s,j));
sf(r)    = sum(i,m0(r,i))-sum(i,e0(r,i));
*ex0(r,i) = sum((s,j),xx0(r,i,s,j))+sum(s,xh0(r,i,s))+sum
(s,xg0(r,i,s))+sum(s,xi0(r,i,s));
ex0(r)   = sum((i,s,j),xx0(r,i,s,j))+sum((i,s),xh0
(r,i,s))+ sum((i,s),xg0(r,i,s))+sum((i,s),xi0(r,i,s));
im0(s)   = sum((r,i,j),xx0(r,i,s,j))+sum
((r,i),xh0(r,i,s))+ sum((r,i),xg0(r,i,s))+sum((r,i),
xi0(r,i,s));


sp0(s)   = fl(s)+fk(s)-sum((r,i),xh0(r,i,s))-td0(s);
ps(s)    = sp0(s)/[fl(s)+fk(s)];
sg0(s)   = td0(s)+sum(j,tz0(s,j))-sum((r,i),xg0(r,i,s));
pg(s)    = sg0(s)/[td0(s)+sum(j,tz0(s,j))];
tr0(s)   = im0(s)-ex0(s);
*tr0(s)  = sp0(s)+sg0(s)+sf(s)-sum((r,i),xi0(r,i,s));


display x0;

*CALIBRATION----------------------------------------------------------------
```

PARAMETER

sig_1	ELASTICITY OF SUBSTITUTION OF FUNCTION1
sig_2	ELASTICITY OF SUBSTITUTION OF FUNCTION2
sig_3	ELASTICITY OF SUBSTITUTION OF FUNCTION3
sig_4	ELASTICITY OF TRANSFORMATION OF FUNCTION4
sig_5	ELASTICITY OF SUBSTITUTION OF FUNCTION5

```
gan_1       SUBSTITUTION ELASTICITY PARAMETER OF FUNCTION 1
gan_2       SUBSTITUTION ELASTICITY PARAMETER OF FUNCTION 2
gan_3       SUBSTITUTION ELASTICITY PARAMETER OF FUNCTION 3
gan_4       TRANSFORMATION ELASTICITY PARAMETER OF FUNCTION 4
gan_5       SUBSTITUTION ELASTICITY PARAMETER OF FUNCTION 5

alp_y(s,j)      SCALE PARAMETER OF FUNCTION1
alp_x(i,s,j)    SCALE PARAMETER OF FUNCTION3
alp_z(r,i)      SCALE PARAMETER OF FUNCTION4
alp_q(r,i)      SCALE PARAMETER OF FUNCTION5

bet_l(s,j)      SHARE PAREMETER OF LABOR
bet_k(s,j)      SHARE PAREMETER OF CAPITAL
bet_x(r,i,s,j)
bet_u(r,i,s)    SHARE PAREMETER OF COMPOSITE HOUSEHOLD
                CONSUMPTION
bet_g(r,i,s)
bet_i(r,i,s)
bet_e(r,i)      SHARE PAREMETER OF EXPORT
bet_d1(r,i)     SHARE PAREMETER OF DOMESTIC
bet_m(r,i)      SHARE PAREMETER OF IMPORT
bet_d2(r,i)     SHARE PAREMETER OF DOMESTIC


ax(i,s,j)
ay(s,j)


tco2(r,i)

sig_z
gan_z
sum_x0(s,j)
```

```
bet_zy(s,j)
bet_zx(i,s,j)
alp_zz(s,j)

sum_x(i,s,j)
;

sig_1  =  2;
sig_2  =  2;
sig_3  =  2;
sig_4  =  2;
sig_5  =  2;

gan_1  =  [sig_1-1]/sig_1;
gan_2  =  [sig_2-1]/sig_2;
gan_3  =  [sig_3-1]/sig_3;
gan_4  =  [sig_4+1]/sig_4;
gan_5  =  [sig_5-1]/sig_5;


bet_l(s,j)$(l0(s,j) and k0(s,j))  =  l0(s,j)**(1-gan_1)/
[l0(s,j)**(1-gan_1)+k0(s,j)**(1-gan_1)];
bet_l(s,j)$(l0(s,j) and not k0(s,j)) = 1;
bet_l(s,j)$(not l0(s,j)) = 0;

bet_k(s,j)$(l0(s,j) and k0(s,j))  =  k0(s,j)**(1-gan_1)/
[l0(s,j)**(1-gan_1)+k0(s,j)**(1-gan_1)];
bet_k(s,j)$(k0(s,j) and not l0(s,j)) = 1;
bet_k(s,j)$(not k0(s,j)) =0;

alp_y(s,j)$(l0(s,j) and k0(s,j))  =  y0(s,j)/{[bet_l(s,j)*
[l0(s,j)**gan_1]+bet_k(s,j)*[k0(s,j)**gan_1]]**(1/gan_1)};
alp_y(s,j)$(l0(s,j) and not k0(s,j))  =  y0(s,j)/l0(s,j);
alp_y(s,j)$(k0(s,j) and not l0(s,j))  =  y0(s,j)/k0(s,j);
```

alp_y(s,j)$(not k0(s,j) and not l0(s,j)) = 0;

sum_x(i,s,j) = sum(w,xx0(w,i,s,j)**(1-gan_2));
bet_x(r,i,s,j)$sum_x(i,s,j) = [xx0(r,i,s,j)**(1-gan_2)]/
sum(w,xx0(w,i,s,j)**(1-gan_2));
bet_x(r,i,s,j)$(not sum_x(i,s,j)) = 0;
alp_x(i,s,j)$x0(i,s,j) = x0(i,s,j)/[[sum(r,bet_x(r,i,s,j)*
[xx0(r,i,s,j)**gan_2])]**(1/gan_2)];
alp_x(i,s,j)$(not x0(i,s,j)) = 0;

bet_u(r,i,s) = xh0(r,i,s)/(fl(s)+fk(s)-td0(s)-sp0(s));
bet_g(r,i,s) = xg0(r,i,s)/(td0(s)+sum(j,tz0(s,j))-sg0(s));
bet_i(r,i,s) = xi0(r,i,s)/(sp0(s)+sg0(s)+sf(s)+tr0(s));

bet_e(r,i)$(e0(r,i) and d0(r,i)) = ([e0(r,i)**(1-gan_4)]/
[e0(r,i)**(1-gan_4)+d0(r,i)**(1-gan_4)]);
bet_e(r,i)$(not e0(r,i)) = 0;
bet_e(r,i)$(e0(r,i) and not d0(r,i)) = 1;

bet_d1(r,i)$(e0(r,i) and d0(r,i)) = ([d0(r,i)**(1-gan_4)]/
[e0(r,i)**(1-gan_4)+d0(r,i)**(1-gan_4)]);
bet_d1(r,i)$(d0(r,i) and not e0(r,i)) = 1;
bet_d1(r,i)$(not d0(r,i)) = 0;

alp_z(r,i)$(e0(r,i) and d0(r,i)) = z0(r,i)/{(bet_e(r,i)*
[e0(r,i)**gan_4]+bet_d1(r,i)*[d0(r,i)**gan_4])**(1/gan_4)}
;
alp_z(r,i)$(e0(r,i) and not d0(r,i)) = z0(r,i)/e0(r,i);
alp_z(r,i)$(d0(r,i) and not e0(r,i)) = z0(r,i)/d0(r,i);
alp_z(r,i)$(not d0(r,i) and not e0(r,i)) = 0;

bet_m(r,i)$(m0(r,i) and d0(r,i)) = [m0(r,i)**(1-gan_5)]/
[m0(r,i)**(1-gan_5)+d0(r,i)**(1-gan_5)];
bet_m(r,i)$(not m0(r,i)) = 0;

```
bet_m(r,i)$(m0(r,i) and not d0(r,i)) = 1;

bet_d2(r,i)$(m0(r,i) and d0(r,i))    = [d0(r,i)**(1-gan_5)]/
[m0(r,i)**(1-gan_5)+d0(r,i)**(1-gan_5)];
bet_d2(r,i)$(d0(r,i) and not m0(r,i)) =   1;
bet_d2(r,i)$(not d0(r,i))   = 0;

alp_q(r,i)$(m0(r,i) and d0(r,i))    = q0(r,i)/{(bet_m(r,i)*
[m0(r,i)**gan_5]+bet_d2(r,i)*[d0(r,i)**gan_5])**(1/gan_5)}
;
alp_q(r,i)$(d0(r,i) and not m0(r,i))    = q0(r,i)/d0(r,i);
alp_q(r,i)$(m0(r,i) and not d0(r,i))    = q0(r,i)/m0(r,i);
alp_q(r,i)$(not m0(r,i) and not d0(r,i)) = 0;

ax(i,s,j)$z0(s,j)  = x0(i,s,j)/z0(s,j);
ay(s,j)$z0(s,j)  =  y0(s,j)/z0(s,j);

tco2(r,i)  =  0;
```

*DEFINE THE MODEL SYSTEM---

VARIABLES
y(s,j)	COMPOSITE FACTOR DEMAND
l(s,j)	LABOR DEMAND
k(s,j)	CAPITAL DEMAND
py(s,j)	PRICE OF Y
pl(s)	PRICE OF L
pk(s)	PRICE OF C
xx(r,i,s,j)	INTERMEDIATE INPUT DEMAND
x(i,s,j)	
pq(r,i)	PRICE OF Q XXH XX XV XG
z(s,j)	OUTPUT
pz(s,j)	PRICE OF Z

```
xh(r,i,s)           COMPOSITE HOUSEHOLD CONSUMPTION DEMAND
xg(r,i,s)
xi(r,i,s)
tz(s,j)             PRODUCTION TAX
tr(s)               TRANSFER
epsilon             EXCHANGE RATE
e(r,i)              EXPORT
d(r,i)              DOMESTIC
pe(r,i)             PRICE OF E IN CNY
pd(r,i)             PRICE OF D
q(r,i)              ARMINGTON
pm(r,i)             PRICE OF M IN CNY
m(r,i)              IMPORT
object
sp(s)
sg(s)
td(s)
tez(r)
teh(r)
*ex(r)
*im(s)
px(i,s,j)

;

EQUATION
eq_y(s,j)
eq_y12(s,j)
eq_y13(s,j)
eq_y14(s,j)
eq_l(s,j)
eq_l2(s,j)
eq_l3(s,j)
eq_k(s,j)
```

```
eq_k2(s,j)
eq_k3(s,j)
eq_x(i,s,j)
eq_x2(i,s,j)
eq_xx(r,i,s,j)
eq_xx2(r,i,s,j)
eq_z(s,j)
eq_y2(s,j)
eq_xz(i,s,j)
eq_xz2(i,s,j)
eq_xh(r,i,s)
eq_xh2(r,i,s)
eq_xg(r,i,s)
eq_xg2(r,i,s)
eq_xi(r,i,s)
eq_xi2(r,i,s)
eq_tz(s,j)
eq_z1(r,i)
eq_z2(r,i)
eq_z3(r,i)
eq_z4(r,i)
eq_e1(r,i)
eq_e2(r,i)
eq_e3(r,i)
eq_d1(r,i)
eq_d2(r,i)
eq_d3(r,i)
eq_q1(r,i)
eq_q2(r,i)
eq_q3(r,i)
eq_q4(r,i)
eq_m1(r,i)
eq_m2(r,i)
eq_m3(r,i)
```

```
eq_d12(r,i)
eq_d22(r,i)
eq_d23(r,i)
eq_td(s)
eq_pe(r,i)
eq_pm(r,i)
eq_mkt(r,i)
eq_fl(s)
eq_fk(s)
eq_sp(s)
eq_sg(s)
*eq_ex(r,i)
*eq_ex(r)
*eq_im(s)
*eq_tr(s)
eq_tr2
eq_sf(s)
eq_tez(r)
eq_teh(r)
eq_y21(s,j)
obj
;
```

eq_y(s,j)$(l0(s,j) and k0(s,j)).. y(s,j) =e= alp_y(s,j)*
[bet_l(s,j)*l(s,j)**gan_1+bet_k(s,j)*k(s,j)**gan_1]**(1/
gan_1);

eq_y12(s,j)$(l0(s,j) and not k0(s,j)).. y(s,j) =e= alp_y
(s,j)*l(s,j);

eq_y13(s,j)$(k0(s,j) and not l0(s,j)).. y(s,j) =e= alp_y
(s,j)*k(s,j);

eq_y14(s,j)$(not y0(s,j)).. y(s,j) =e= 0;

```
eq_l(s,j)$(l0(s,j)and k0(s,j))..  l(s,j)*alp_y(s,j)  =e=  y(s,j)
*{py(s,j)*alp_y(s,j)*bet_l(s,j)/pl(s)}**[1/(1-gan_1)];

eq_l2(s,j)$(l0(s,j) and not k0(s,j))..  l(s,j)*pl(s)  =e=  y(s,j)
*py(s,j);

eq_l3(s,j)$(not l0(s,j))..  l(s,j)  =e=  0;

eq_k(s,j)$(l0(s,j)and k0(s,j))..  k(s,j)*alp_y(s,j)  =e=  y(s,j)*
{py(s,j)*alp_y(s,j)*bet_k(s,j)/pk(s)}**[1/(1-gan_1)];

eq_k2(s,j)$(k0(s,j) and not l0(s,j))..  k(s,j)*pk(s)  =e=  y(s,j)
*py(s,j);

eq_k3(s,j)$(not k0(s,j))..  k(s,j)  =e=  0;

eq_x(i,s,j)$x0(i,s,j)..  x(i,s,j)  =e=  alp_x(i,s,j)*
[sum(r,bet_x(r,i,s,j)*xx(r,i,s,j)**gan_2)]**(1/gan_2);

eq_x2(i,s,j)$(not x0(i,s,j))..  x(i,s,j)  =e=  0;

eq_xx(r,i,s,j)$xx0(r,i,s,j)..  xx(r,i,s,j)*alp_x(i,s,j)  =e=
x(i,s,j)*{px(i,s,j)*alp_x(i,s,j)*bet_x(r,i,s,j)/[tco2(r,i)
+pq(r,i)]}**(1/[1-gan_2]);

eq_xx2(r,i,s,j)$(not xx0(r,i,s,j))..  xx(r,i,s,j)  =e=  0;

eq_y2(s,j)$y0(s,j)..  y(s,j)  =e=  ay(s,j)*z(s,j);

eq_y21(s,j)$(not y0(s,j))..  y(s,j)  =e=  0;

eq_xz(i,s,j)$(x0(i,s,j)  gt  0)..  x(i,s,j)  =e=  ax(i,s,j)
*z(s,j);
```

```
eq_xz2(i,s,j)$(x0(i,s,j) = 0)..  x(i,s,j) =e= 0;

eq_z(s,j)..    pz(s,j) =e= py(s,j)*ay(s,j)+sum(i,px(i,s,j)*
ax(i,s,j));

*eq_xh(r,i,s)$(xh0(r,i,s)  gt  0)..    {[pq(r,i)+tco2*co2(i)*
sss(i)]**0.5}*xh(r,i,s)*sum[(w,h),[(pq(w,h)+tco2*co2(h)*ss
s(h))**0.5]*bet_u(w,h,s)]   =e=   bet_u(r,i,s)*[pl(s)*fl(s)+
pk(s)*fk(s)-td(s)-sp(s)];
eq_xh(r,i,s)$(xh0(r,i,s)   gt   0)..      [pq(r,i)+tco2(r,i)]*xh
(r,i,s)  =e=  bet_u(r,i,s)*[pl(s)*fl(s)+pk(s)*fk(s)-td(s)-sp(s)];
*eq_xh(r,i,s)$(xh0(r,i,s)   gt   0)..    xh(r,i,s)*pq(r,i)   =e=
bet_u(r,i,s)*[pl(s)*fl(s)+pk(s)*fk(s)-td(s)-sp(s)];

eq_xh2(r,i,s)$(xh0(r,i,s) = 0)..  xh(r,i,s) =e= 0;

eq_xg(r,i,s)$(xg0(r,i,s)    gt    0)..      [pq(r,i)+tco2(r,i)]*xg
(r,i,s) =e= bet_g(r,i,s)*[td(s)+sum(j,tz(s,j))+tez(s)+teh(s)-sg(s)];

*eq_xg(r,i,s)$(xg0(r,i,s) gt 0).. xg(r,i,s)*pq(r,i)  =e=  bet_g
(r,i,s)*[td(s)+sum(j,tz(s,j))-sg(s)];

eq_xg2(r,i,s)$(xg0(r,i,s) = 0)..  xg(r,i,s) =e= 0;

eq_xi(r,i,s)$(xi0(r,i,s)    gt    0)..     [pq(r,i)+tco2(r,i)]
*xi(r,i,s)  =e=  bet_i(r,i,s)*[sp(s)+sg(s)+tr(s)+epsilon*sf(s)];

*eq_xi(r,i,s)$(xi0(r,i,s)   gt   0)..     xi(r,i,s)*pq(r,i)   =e=
bet_i(r,i,s)*[sp(s)+sg(s)+tr(s)+epsilon*sf(s)];

eq_xi2(r,i,s)$(xi0(r,i,s) = 0)..  xi(r,i,s)  =e=  0;

eq_tz(s,j)..   tz(s,j) =e= tauz(s,j)*pz(s,j)*z(s,j);
```

```
eq_z1(r,i)$(e0(r,i) and d0(r,i))..  z(r,i) =e= alp_z(r,i)*
[bet_e(r,i)*[e(r,i)**gan_4]+bet_d1(r,i)*[d(r,i)**gan_4]]**
{1/gan_4};

eq_z2(r,i)$(d0(r,i) and not e0(r,i))..   z(r,i) =e= alp_z
(r,i)*d(r,i);

eq_z3(r,i)$(e0(r,i) and not d0(r,i))..   z(r,i) =e= alp_z
(r,i)*e(r,i);

eq_z4(r,i)$(not z0(r,i))..  z(r,i) =e= 0;

eq_e1(r,i)$(e0(r,i) and d0(r,i))..  e(r,i)*alp_z(r,i) =e= z
(r,i)*{{(1+tauz(r,i))*pz(r,i)*alp_z(r,i)*bet_e(r,i)/pe(r,i
)}**[1/(1-gan_4)]};

eq_e2(r,i)$(e0(r,i) and not d0(r,i))..  pe(r,i)*e(r,i) =e=
(1+tauz(r,i))*pz(r,i)*z(r,i);

eq_e3(r,i)$(e0(r,i) = 0)..  e(r,i) =e= 0;

eq_d1(r,i)$(d0(r,i) and e0(r,i))..   d(r,i)*alp_z(r,i) =e=
z(r,i)*{{(1+tauz(r,i))*pz(r,i)*alp_z(r,i)*bet_d1(r,i)/pd(r
,i)}**[1/(1-gan_4)]};

eq_d2(r,i)$(d0(r,i) and not e0(r,i))..  pd(r,i)*d(r,i) =e=
(1+tauz(r,i))*pz(r,i)*z(r,i);

eq_d3(r,i)$(d0(r,i) = 0)..  d(r,i) =e= 0;

eq_q1(r,i)$(m0(r,i) and d0(r,i))..  q(r,i) =e= alp_q(r,i)
*[bet_m(r,i)*m(r,i)**gan_5+bet_d2(r,i)*d(r,i)**gan_5]**{1/
gan_5};
```

```
eq_q2(r,i)$(d0(r,i) and not m0(r,i))..    q(r,i)  =e=  alp_q
(r,i)*d(r,i);

eq_q3(r,i)$(m0(r,i) and not d0(r,i))..    q(r,i)  =e=  alp_q
(r,i)*m(r,i);

eq_q4(r,i)$(not q0(r,i))..   q(r,i)  =e=  0;

eq_m1(r,i)$(m0(r,i) and d0(r,i))..   m(r,i)*alp_q(r,i)  =e=
q(r,i)*{{pq(r,i)*alp_q(r,i)*bet_m(r,i)/pm(r,i)}**[1/(1-ga
n_5)]};

eq_m2(r,i)$(not m0(r,i))..   m(r,i)  =e=  0;

eq_m3(r,i)$(m0(r,i) and not d0(r,i))..   pm(r,i)*m(r,i)  =e=
pq(r,i)*q(r,i);

eq_d12(r,i)$(d0(r,i) and m0(r,i))..   d(r,i)*alp_q(r,i)  =e=
q(r,i)*{{pq(r,i)*alp_q(r,i)*bet_d2(r,i)/pd(r,i)}**{1/(1-ga
n_5)}};

eq_d22(r,i)$(d0(r,i) and not m0(r,i))..   pd(r,i)*d(r,i)  =e=
pq(r,i)*q(r,i);

eq_d23(r,i)$(not d0(r,i))..   d(r,i)  =e=  0;

eq_sf(r)..    sf(r)  =e=  sum((i),m(r,i))-sum((i),e(r,i));

eq_td(s)..    td(s)  =e=  taud*[pl(s)*fl(s) +pk(s)*fk(s)];

eq_pe(r,i)..    pe(r,i)  =e=  epsilon;

eq_pm(r,i)..    pm(r,i)  =e=  epsilon;
```

```
eq_mkt(r,i)..    q(r,i)   =e=  sum(s,xh(r,i,s))+sum((s,j),xx
(r,i,s,j))+sum(s,xg(r,i,s))+sum(s,xi(r,i,s));

eq_fl(s)..  sum(j,l(s,j)) =e= fl(s);

eq_fk(s)..  sum(j,k(s,j)) =e= fk(s);

eq_sp(s)..  sp(s) =e= ps(s)*[pl(s)*fl(s)+pk(s)*fk(s)];

eq_sg(s)..   sg(s)   =e=  pg(s)*[td(s)+sum(j,tz(s,j))+tez(s)
+teh(s)];

eq_tez(s)..  tez(s) =e= sum((r,i,j),tco2(r,i)*xx(r,i,s,j));

eq_teh(s)..  teh(s) =e= sum((r,i),tco2(r,i)*[xh(r,i,s) +xg
(r,i,s)+xi(r,i,s)]);

*eq_tr(s)..  tr(s) =e= sp(s)+sg(s)+epsilon*sf(s)-sum((r,i),
pq(r,i)*xi(r,i,s));

eq_tr2..  sum(s,tr(s)) =e= 0;

obj..  object =e= epsilon;
```

*INITIALIZING VARIABLES--

```
y.l(s,j)    = y0(s,j);
l.l(s,j)    = l0(s,j);
k.l(s,j)    = k0(s,j);
xx.l(r,i,s,j)  = xx0(r,i,s,j);
x.l(i,s,j)  = x0(i,s,j);
z.l(s,j)    = z0(s,j);
xh.l(r,i,s) = xh0(r,i,s);
```

```
tz.l(s,j)    = tz0(s,j);
epsilon.l    = 1;
e.l(r,i)     = e0(r,i);
d.l(r,i)     = d0(r,i);
q.l(r,i)     = q0(r,i);
m.l(r,i)     = m0(r,i);
py.l(s,j)    = 1;
pl.l(s)      = 1;
pk.l(s)      = 1;
px.l(i,s,j)  = 1;
pq.l(r,i)    = 1;
pz.l(s,j)    = 1;
pe.l(r,i)    = 1;
pd.l(r,i)    = 1;
pm.l(r,i)    = 1;
tr.l(s)      = tr0(s);
td.l(s)      = td0(s);
sp.l(s)      = sp0(s);
sg.l(s)      = sg0(s);
*ex.l(r)     = ex0(r);
*im.l(s)     = im0(s);
xg.l(r,i,s)  = xg0(r,i,s);
xi.l(r,i,s)  = xi0(r,i,s);
*tez.l(r)    = tez0(r);
*teh.l(r)    = teh0(r);
```

*SETTING LOWER BOUNDS TO AVOID DIVISION BY ZERO-------------------------------

```
y.lo(s,j)      = eps;
l.lo(s,j)      = eps;
k.lo(s,j)      = eps;
xx.lo(r,i,s,j) = 0;
```

```
x.lo(i,s,j)    = 0;
z.lo(s,j)      = eps;
xh.lo(r,i,s)   = eps;
epsilon.lo     = eps;
e.lo(r,i)      = eps;
d.lo(r,i)      = eps;
q.lo(r,i)      = eps;
m.lo(r,i)      = eps;
py.lo(s,j)     = eps;
pl.lo(s)       = eps;
pk.lo(s)       = eps;
px.lo(i,s,j)   = eps;
pq.lo(r,i)     = eps;
pz.lo(s,j)     = eps;
pe.lo(r,i)     = eps;
pd.lo(r,i)     = eps;
pm.lo(r,i)     = eps;
tez.lo(r)      = eps;
teh.lo(r)      = eps;
xg.lo(r,i,s)   = eps;
xi.lo(r,i,s)   = eps;

sp.lo(s)       = eps;

*ex.lo(r)      = eps;
*im.lo(s)      = eps;




epsilon.fx  = 1;
```

*DEFINING AND SOLVE THE MODEL--

```
MODEL project23 /ALL/;
project23.iterlim = 10000;
SOLVE project23 MAXIMIZING object USING NLP;
```

Model 2: Multi-National, Multi-Region, Static Model

Codes based on GAMS 23.0 Software for Models
Multi Nations Multi-Regional Model BAU Solution:
$TITLE WORLD MULTI REGION CGE MODEL

*DEFINITION OF SETS FOR SUFFIX

```
set u/ocn,jpn,gca,roa,ind,usa,rus,eu27,row,
l,k,pt,dt,hoh,gov,inv,ex,im,la,ene,lat,lt,kt,enet,ct,exct,
imct,dtax,hsaving,gsaving,
agr,cm,ogm,om,fm,tex,wa,swp,pp,ppc,ci,nmmp,msp,mp,mi,te,em
e,ece,omi,ewgs,cons,tw,com,ser/

r/noea,nomu,noco,eaco,soco,cent,nowe,sowe/

i/agr,cm,ogm,om,fm,tex,wa,swp,pp,ppc,ci,nmmp,msp,mp,mi,te,
eme,ece,omi,ewgs,cons,tw,com,ser/

rr      Out of China
/ocn,jpn,gca,roa,ind,usa,rus,eu27,row/;

*OCN        Oceania
*JPN        Japan
*GCA        Hongkong,Macau and Taiwan
```

Appendix *131*

*ROA	Rest of Asia
*IND	India
*USA	USA
*RUS	Russia
*EU27	EU27
*ROW	Rest of the world

```
alias (u,v),(i,j,h),(r,s,w,vv),(rr,ss,ww);
```

*LOADING DATA
```
table i_m(r,i,s,j)
$INCLUDE in111.INC
;

table f_d(r,i,s,v)
$INCLUDE fd111.INC
;

table v_a(u,s,j)
$INCLUDE vd111.INC
;

TABLE EXT(R,I,V)
$INCLUDE imex111.INC
;

table i_mw(rr,i,j)
$INCLUDE inw111.INC
;

table v_dw(rr,u,j)
$INCLUDE vdw111.INC

table f_dw(rr,i,v)
```

```
$INCLUDE fdw111.INC
;

table i_mtw(rr,i,j)
$INCLUDE intw111.INC
;

TABLE WEX(rr,ss,i)
$INCLUDE exw1111.INC
;

TABLE WEXT(rr,ss,i)
$INCLUDE extw1111.INC
;

TABLE WIM(rr,ss,i)
$INCLUDE imw1111.INC
;

TABLE WIMT(rr,ss,i)
$INCLUDE imtw1111.INC
;

TABLE WEXIMC(rr,i,v)
$INCLUDE eximc111.INC
;

TABLE SAVING(rr,v)
$INCLUDE sav111.inc
;

table enec(i,j)
$INCLUDE enec1111.inc
```

;

```
table enew(rr,i,j)
$INCLUDE enew1111.inc
;

table enewh(rr,i,v)
$INCLUDE enewh1111.inc
;

table enech(i,v)
$INCLUDE enech1111.inc
;
```

*LOADING THE INITIAL VALUE
```
parameter
eneww(rr,i,j)
enecc(i,j)
enewhh(rr,i)
enechh(i)
eneccc(r,i,s,j)
enechhh(r,i,s)
beneww(rr,i,j)
benewhh(rr,i)
beneccc(r,i,s,j)
benechhh(r,i,s)
y0(s,j)              COMPOSITE FACTOR
l0(s,j)              LABOR
k0(s,j)              CAPITAL
xx0(r,i,s,j)         INTERMEDIATE INPUT
x0(i,s,j)            COMPOSITE INERTMEDIATE INPUT
z0(s,j)              OUTPUT
xh0(r,i,s)           HOUSEHOLD CONSUMPTION
xg0(r,i,s)           GOVERNMENT CONSUMPTION
```

xi0(r,i,s)	INVESTMENT
tz0(s,j)	PRODUCTION TAX
epsilon0	EXCHANGE RATE
e0(r,i)	EXPORT
m0(r,i)	IMPORT
d0(r,i)	DOMESTIC
q0(r,i)	ARMINGTON
tauz(s,j)	PRODUCTION TAX
taud	PRODUCTION TAX RATE
sf(r)	FOREIGN SAVING IN US DOLLAR
fl(s)	FACTOR ENDOWMENT OF LABOR
fk(s)	FACTOR ENDOWMENT OF CAPITAL
tr0(s)	TRANSFER
ex0(r)	CAPITAL OUTPUT FLOW OF REGION R
im0(s)	CAPITAL INPUT FLOW OF REGION R
sp0(s)	PRAVITE SAVING
sg0(s)	GOVERNMENT SAVING
td0(s)	DICRET TAX
ps(s)	PRAVITE SAVING RATE
pg(s)	GOVERNMENT SAVING RATE
tco2(r,i)	

wx0(rr,i,j)	INTERMEDIATE INPUT i TO j SECTOR OF REGION rr
wlab0(rr,j)	LABOR INPUT TO j SECTOR OF REGION rr
wk0(rr,j)	CAPITAL INPUT TO j SECTOR OF REGION rr
wlan0(rr,j)	LAND INPUT TO J SECTOR OF REGION RR
wene0(rr,j)	NATURE RESOUCE INPUT TO J SECTOR OF REGION RR
wy0(rr,j)	COMPOSITE VALUE ADDED INPUT TO J SECTOR OF REGION RR
wtlab0(rr,j)	LABOR INPUT TAX OF J SECTOR IN REGION RR
wtaulab(rr,j)	LABOR INPUT TAX RATE OF J SECTOR IN REGION RR
wtk0(rr,j)	CAPITAL INPUT TAX OF J SECTOR IN REGION RR
wtauk(rr,j)	CAPITAL INPUT TAX RATE OF J SECTOR IN REGION RR

wtlan0(rr,j)	LAND INPUT TAX OF J SECTOR IN REGION RR
wtaulan(rr,j)	LAND INPUT TAX RATE OF J SECTOR IN REGION RR
wtene0(rr,j)	NATURE RESOUCE INPUT TAX OF J SECTOR IN REGION RR
wtauene(rr,j)	NATURE RESOURCE INPUT TAX RATE OF J SECTOR IN REGION RR
wtx0(rr,i,j)	INTERMEDIATE I'S INPUT TAX OF J SECTOR IN REGION RR
wtaux(rr,i,j)	INTERMEDIATE I'S INPUT TAX RATE OF J SECTOR IN REGION RR
wtz0(rr,j)	SECTOR J'S PRODUCTION TAX OF REGION RR
wtauz(rr,j)	SECTOR J'S PRODUCTION TAX RATE OF REGION RR
wtd0(rr)	DIRECT TAX IN REGION RR
wtaud(rr)	DIRECT TAX RATE OF REGION RR
wxh0(rr,i)	HOUSEHOLD CONSUMPTION ON COMMDITY I IN REGION RR
wxg0(rr,i)	GOVERNMENT CONSUMPTION ON COMMDITY I IN REGION RR
wxi0(rr,i)	INVESTMENT CONSUMPTION ON COMMDITY I IN REGION RR
wd0(rr,i)	DOMESTIC SUPPLY OF GOODS I IN REGION RR
wz0(rr,j)	TOTAL PRODUCTION OF COMMDITY I IN REGION RR
wq0(rr,i)	COMPOSITE GOODS I BY DOMESTIC AND IMPORT IN REGION RR
wth0(rr,i)	CONSUMPTION TAX OF COMMDITY I IN REGION RR
wtauh(rr,i)	CONSUMPTION TAX RATE OF COMMDITY I IN REGION RR
wee0(rr,ss,i)	EXPORT GOODS I FROM RR TO SS
wmm0(rr,ss,i)	IMPORT GOODS I FROM SS TO RR
ej0(rr,i)	RR'S EXPORT GOODS I TO cHINA
mj0(rr,i)	RR'S IMPORT GOODS I FROM CHINA
we0(rr,i)	COMPOSITE EXPORT GOODS I FROM REGION RR
wm0(rr,i)	
wte0(rr,ss,i)	
wtaue(rr,ss,i)	
wtej0(rr,i)	

```
wtauej(rr,i)
wtm0(rr,ss,i)
wtaum(rr,ss,i)
wtmj0(rr,i)
wtaumj(rr,i)
wsf(rr)
wflab(rr)
wflan(rr)
wfk(rr)
wfene(rr)
wsp0(rr)
wsg0(rr)
wps(rr)
wpg(rr)
wepsilon0(rr)
```

sig_1	ELASTICITY OF SUBSTITUTION OF FUNCTION1
sig_2	ELASTICITY OF SUBSTITUTION OF FUNCTION2
sig_3	ELASTICITY OF SUBSTITUTION OF FUNCTION3
sig_4	ELASTICITY OF TRANSFORMATION OF FUNCTION4
sig_5	ELASTICITY OF SUBSTITUTION OF FUNCTION5
gan_1	SUBSTITUTION ELASTICITY PARAMETER OF FUNCTION2
gan_2	SUBSTITUTION ELASTICITY PARAMETER OF FUNCTION2
gan_3	SUBSTITUTION ELASTICITY PARAMETER OF FUNCTION3
gan_4	TRANSFORMATION ELASTICITY PARAMETER OF FUNCTION4
gan_5	SUBSTITUTION ELASTICITY PARAMETER OF FUNCTION5
alp_y(s,j)	SCALE PARAMETER OF FUNCTION1
alp_x(i,s,j)	SCALE PARAMETER OF FUNCTION3
alp_z(r,i)	SCALE PARAMETER OF FUNCTION4
alp_q(r,i)	SCALE PARAMETER OF FUNCTION5

bet_l(s,j)	SHARE PAREMETER OF LABOR
bet_k(s,j)	SHARE PAREMETER OF CAPITAL
bet_x(r,i,s,j)	
bet_u(r,i,s)	SHARE PAREMETER OF COMPOSITE HOUSEHOLD CONSUMPTION
bet_g(r,i,s)	
bet_i(r,i,s)	
bet_e(r,i)	SHARE PAREMETER OF EXPORT
bet_d1(r,i)	SHARE PAREMETER OF DOMESTIC
bet_m(r,i)	SHARE PAREMETER OF IMPORT
bet_d2(r,i)	SHARE PAREMETER OF DOMESTIC

```
tco2(r,i)
twco2(rr,i)
ax(i,s,j)
ay(s,j)
sig_z
gan_z
sum_x0(s,j)
bet_zy(s,j)
bet_zx(i,s,j)
alp_zz(s,j)

sum_x(i,s,j)

sig_w1
sig_w4
sig_w5
sig_wm
sig_we
gan_w1
gan_w4
gan_w5
gan_wm
```

```
gan_we
alp_wy(rr,j)
alp_wz(rr,i)
alp_wee(rr,i)
alp_wmm(rr,i)
alp_wq(rr,i)
bet_wlab(rr,j)
bet_wlan(rr,j)
bet_wk(rr,j)
bet_wene(rr,j)
bet_wxh(rr,i)
bet_wxg(rr,i)
bet_wxi(rr,i)
bet_wee(rr,ss,i)
bet_ej(rr,i)
bet_we(rr,i)
bet_wd(rr,i)
bet_wmm(rr,ss,i)
bet_mj(rr,i)
bet_wm(rr,i)
bet_wd2(rr,i)
awx(rr,i,j)
awy(rr,j)
;

eneww(rr,i,j)   = enew(rr,i,j);
enecc(i,j)      = enec(i,j);
enewhh(rr,i)    = enewh(rr,i,'hoh');
enechh(i)       = enech(i,'hoh');
l0(s,j)         = V_A("L",s,j);
k0(s,j)         = V_A("K",s,j);
tz0(s,j)        = V_A("PT",s,j);
y0(s,j)         = l0(s,j)+k0(s,j);
xx0(r,i,s,j)    = I_M(r,i,s,j);
```

```
x0(i,s,j)    = sum(r,xx0(r,i,s,j));
eneccc(r,i,s,j) = enecc(i,j)*[xx0(r,i,s,j)/x0(i,s,j)];
z0(s,j)      = y0(s,j)+sum(i,x0(i,s,j));
xh0(r,i,s)   = F_D(r,i,s,"hoh");
enechhh(r,i,s) $ xh0(r,i,s) = enechh(i)*(xh0(r,i,s)/sum(w,
xh0(w,i,s)));
enechhh(r,i,s) $ (not xh0(r,i,s)) = 0;
xg0(r,i,s)   = F_D(r,i,s,"gov");
xi0(r,i,s)   = F_D(r,i,s,"inv");
fl(s)        = sum(j,l0(s,j));
fk(s)        = sum(j,k0(s,j));
tauz(s,j)$(z0(s,j))  = tz0(s,j)/z0(s,j);
taud = 0.1;
td0(s)       = taud*[fl(s)+fk(s)];
epsilon0 = 1;
e0(r,i)      = ext(r,i,"ex");
m0(r,i)      = (-1)*EXT(r,i,"im");
d0(r,i)      = (1+tauz(r,i))*z0(r,i)-e0(r,i);
q0(r,i)      = sum(s,xh0(r,i,s))+sum(s,xg0(r,i,s))+sum(s,xi0(r,
i,s))+sum((s,j),xx0(r,i,s,j));
sf(r)        = sum(i,m0(r,i))-sum(i,e0(r,i));
ex0(r)       = sum((i,s,j),xx0(r,i,s,j))+sum((i,s),xh0 (r,i,s))+
sum((i,s),xg0(r,i,s))+sum((i,s),xi0(r,i,s));
im0(s)       = sum((r,i,j),xx0(r,i,s,j))+sum((r,i),xh0 (r,i,s))+
sum((r,i),xg0(r,i,s))+sum((r,i),xi0(r,i,s));
sp0(s)       = fl(s)+fk(s)-sum((r,i),xh0(r,i,s))-td0(s);
ps(s)        = sp0(s)/[fl(s)+fk(s)];
sg0(s)       = td0(s)+sum(j,tz0(s,j))-sum((r,i),xg0(r,i,s));
pg(s)        = sg0(s)/[td0(s)+sum(j,tz0(s,j))];
tr0(s)       = im0(s)-ex0(s);


wlab0(rr,j)  = v_dw(rr,'l',j);
wk0(rr,j)    = v_dw(rr,'k',j);
```

```
wlan0(rr,j)   =  v_dw(rr,'la',j);
wene0(rr,j)   =  v_dw(rr,'ene',j);
wtz0(rr,j)    =  v_dw(rr,'pt',j);
wtlab0(rr,j)  =  v_dw(rr,'lt',j);
wtaulab(rr,j) = wtlab0(rr,j)/wlab0(rr,j);
wtk0(rr,j)    =  v_dw(rr,'kt',j);
wtauk(rr,j)   = wtk0(rr,j)/wk0(rr,j);
wtlan0(rr,j)  =  v_dw(rr,'lat',j);
wtaulan(rr,j)$(wlan0(rr,j))  = wtlan0(rr,j)/wlan0(rr,j);
wtene0(rr,j)  =  v_dw(rr,'enet',j);
wtauene(rr,j)$(wene0(rr,j))  = wtene0(rr,j)/wene0(rr,j);
wy0(rr,j)   =  (1+wtaulab(rr,j))*wlab0(rr,j)+(1+wtauk (rr,j))
*wk0(rr,j)+(1+wtaulan(rr,j))*wlan0(rr,j)+(1+wtauene(rr,j))
*wene0(rr,j);
wx0(rr,i,j)   =  i_mw(rr,i,j);
wtx0(rr,i,j)  =  i_mtw(rr,i,j);
wtaux(rr,i,j) =  wtx0(rr,i,j)/wx0(rr,i,j);
wz0(rr,j) = wy0(rr,j) + sum(i,(1+wtaux(rr,i,j)) *wx0 (rr,i,j));
wxh0(rr,i)  =   f_dw(rr,i,'hoh');
wxg0(rr,i)  =   f_dw(rr,i,'gov');
wxi0(rr,i)  =   f_dw(rr,i,'inv');
wflab(rr)   = sum(j,wlab0(rr,j));
wfk(rr)    = sum(j,wk0(rr,j));
wflan(rr)   = sum(j,wlan0(rr,j));
wfene(rr)   = sum(j,wene0(rr,j));
wtauz(rr,j)$(wz0(rr,j))   = wtz0(rr,j)/wz0(rr,j);
wtd0(rr)   = SAVING(rr,"dtax");
wtaud(rr)  = wtd0(rr)/[wflab(rr)+wfk(rr)+wflan(rr) +wfene (rr)];
wth0(rr,i)  =   f_dw(rr,i,'ct');
wtauh(rr,i) =  wth0(rr,i)/wxh0(rr,i);
wepsilon0(rr)   = 1;
wee0(rr,ss,i)   =   WEX(rr,ss,i);
wmm0(rr,ss,i)   =   WIM(rr,ss,i);
ej0(rr,i)   =   WEXIMC(rr,i,'ex');
```

```
mj0(rr,i)   =  WEXIMC(rr,i,'im');
wte0(rr,ss,i)  =  WEXT(rr,ss,i);
wtaue(rr,ss,i)$(wee0(rr,ss,i))  =wte0(rr,ss,i)/wee0(rr,ss,i);
wtej0(rr,i)  =  WEXIMC(rr,i,'exct');
wtauej(rr,i)  = wtej0(rr,i)/ej0(rr,i);
we0(rr,i)  = sum(ss,(1-wtaue(rr,ss,i))*wee0(rr,ss,i)) +(1-
wtauej(rr,i))*ej0(rr,i);
wtm0(rr,ss,i)  =  WIMT(rr,ss,i);
wtaum(rr,ss,i)$(wmm0(rr,ss,i)) = wtm0(rr,ss,i)/wmm0(rr,ss,i);
wtmj0(rr,i)  =  WEXIMC(rr,i,'imct');
wtaumj(rr,i)$(mj0(rr,i))  =  wtmj0(rr,i)/mj0(rr,i);
wm0(rr,i)  = sum(ss,(1+wtaum(rr,ss,i))*wmm0(rr,ss,i)) +(1+
wtaumj(rr,i))*mj0(rr,i);
wd0(rr,i)  = (1+wtauz(rr,i))*wz0(rr,i)-we0(rr,i);
wq0(rr,i) = wxh0(rr,i)+wxg0(rr,i)+wxi0(rr,i)+sum(j,wx0(rr,i,j));
wsf(rr)  = sum((ss,i),wmm0(rr,ss,i))+sum(i,mj0(rr,i)) -sum
((ss,i),wee0(rr,ss,i))-sum(i,ej0(rr,i));
wsp0(rr)  = saving(rr,"hsaving");
wps(rr)  = wsp0(rr)/[wflab(rr)+wfk(rr)+wflan(rr) +wfene(rr)];
wsg0(rr)  = saving(rr,"gsaving");
wpg(rr)  = wsg0(rr)/[wtd0(rr)+sum(j,wtz0(rr,j)) +sum (i,wth0
(rr,i))+sum((i,j),wtx0(rr,i,j))+sum(j,wtlab0(rr,j))
+sum(j,wtk0(rr,j))+sum(j,wtlan0(rr,j))+sum(j,wtene0(rr,j))
+sum((ss,i),wte0(rr,ss,i))+sum(i,wtej0(rr,i))
+sum((ss,i),wtm0(rr,ss,i))+sum(i,wtmj0(rr,i))];

beneww(rr,i,j)  = eneww(rr,i,j)/wx0(rr,i,j);
benewhh(rr,i)  = enewhh(rr,i)/wxh0(rr,i);
beneccc(r,i,s,j)  = eneccc(r,i,s,j)/xx0(r,i,s,j);
benechhh(r,i,s)$ xh0(r,i,s)  = enechhh(r,i,s)/xh0(r,i,s);
benechhh(r,i,s)$(not xh0(r,i,s))  = 0;

*CALIBRATION----------------------------------------------------------
sig_1 =2;
```

```
sig_2   =   2;
sig_3   =   2;
sig_4   =   2;
sig_5   =   2;

gan_1   =   [sig_1-1]/sig_1;
gan_2   =   [sig_2-1]/sig_2;
gan_3   =   [sig_3-1]/sig_3;
gan_4   =   [sig_4+1]/sig_4;
gan_5   =   [sig_5-1]/sig_5;

bet_l(s,j)$(l0(s,j)   and   k0(s,j))   =   l0(s,j)**(1-gan_1)/
[l0(s,j)**(1-gan_1)+k0(s,j)**(1-gan_1)];
bet_l(s,j)$(l0(s,j) and not k0(s,j)) = 1;
bet_l(s,j)$(not l0(s,j)) = 0;

bet_k(s,j)$(l0(s,j)   and   k0(s,j))   =   k0(s,j)**(1-gan_1)/
[l0(s,j)**(1-gan_1)+k0(s,j)**(1-gan_1)];
bet_k(s,j)$(k0(s,j) and not l0(s,j)) = 1;
bet_k(s,j)$(not k0(s,j)) = 0;

alp_y(s,j)$(l0(s,j) and k0(s,j))   = y0(s,j)/{[bet_l(s,j)*
[l0(s,j)**gan_1]+bet_k(s,j)*[k0(s,j)**gan_1]]**(1/gan_1)};
alp_y(s,j)$(l0(s,j) and not k0(s,j)) = y0(s,j)/l0(s,j);
alp_y(s,j)$(k0(s,j) and not l0(s,j)) = y0(s,j)/k0(s,j);
alp_y(s,j)$(not k0(s,j) and not l0(s,j)) = 0;

bet_x(r,i,s,j)$(sum(w,xx0(w,i,s,j)**(1-gan_2))) = [xx0 (r,i,s,j)
**(1-gan_2)]/sum(w,xx0(w,i,s,j)**(1-gan_2));

alp_x(i,s,j)$x0(i,s,j)   =   x0(i,s,j)/[[sum(r,bet_x(r,i,s,j)
*[xx0(r,i,s,j)**gan_2])]**(1/gan_2)];
alp_x(i,s,j)$(not x0(i,s,j)) = 0;
```

```
bet_u(r,i,s)   = xh0(r,i,s)/(fl(s)+fk(s)-td0(s)-sp0(s));
bet_g(r,i,s)   = xg0(r,i,s)/(td0(s)+sum(j,tz0(s,j))-sg0(s));
bet_i(r,i,s)   = xi0(r,i,s)/(sp0(s)+sg0(s)+sf(s)+tr0(s));


bet_e(r,i)$(e0(r,i) and d0(r,i))   = ([e0(r,i)**(1-gan_4)]/
[e0(r,i)**(1-gan_4)+d0(r,i)**(1-gan_4)]);
bet_e(r,i)$(not e0(r,i))   = 0;
bet_e(r,i)$(e0(r,i) and not d0(r,i))   = 1;

bet_d1(r,i)$(e0(r,i) and d0(r,i))   = ([d0(r,i)**(1-gan_4)]
/[e0(r,i)**(1-gan_4)+d0(r,i)**(1-gan_4)]);
bet_d1(r,i)$(d0(r,i) and not e0(r,i))   = 1;
bet_d1(r,i)$(not d0(r,i))   = 0;


alp_z(r,i)$(e0(r,i) and d0(r,i))   = z0(r,i)/{(bet_e(r,i)*
[e0(r,i)**gan_4]+bet_d1(r,i)*[d0(r,i)**gan_4])**(1/gan_4)};
alp_z(r,i)$(e0(r,i) and not d0(r,i))   = z0(r,i)/e0(r,i);
alp_z(r,i)$(d0(r,i) and not e0(r,i))   = z0(r,i)/d0(r,i);
alp_z(r,i)$(not d0(r,i) and not e0(r,i))   = 0;


bet_m(r,i)$(m0(r,i) and d0(r,i))   = [m0(r,i)**(1-gan_5)]/[m0
(r,i)**(1-gan_5)+d0(r,i)**(1-gan_5)];
bet_m(r,i)$(not m0(r,i))   = 0;
bet_m(r,i)$(m0(r,i) and not d0(r,i))   = 1;


bet_d2(r,i)$(m0(r,i) and d0(r,i))   = [d0(r,i)**(1-gan_5)]/
[m0(r,i)**(1-gan_5)+d0(r,i)**(1-gan_5)];
bet_d2(r,i)$(d0(r,i) and not m0(r,i))   = 1;
bet_d2(r,i)$(not d0(r,i))   = 0;


alp_q(r,i)$(m0(r,i) and d0(r,i))   = q0(r,i)/{(bet_m(r,i)*
[m0(r,i)**gan_5]+bet_d2(r,i)*[d0(r,i)**gan_5])**(1/gan_5)};
alp_q(r,i)$(d0(r,i) and not m0(r,i))   = q0(r,i)/d0(r,i);
alp_q(r,i)$(m0(r,i) and not d0(r,i))   = q0(r,i)/m0(r,i);
```

```
alp_q(r,i)$(not m0(r,i) and not d0(r,i))  = 0;

ax(i,s,j)$z0(s,j)  = x0(i,s,j)/z0(s,j);
ay(s,j)$z0(s,j)  = y0(s,j)/z0(s,j);
twco2(rr,i)  =  0;
tco2(r,i)  =  0;


sig_w1  =  2;
sig_w4  =  2;
sig_w5  =  2;
sig_wm  =  2;
sig_we  =  2;
gan_w1  =  [sig_w1-1]/sig_w1;
gan_w4  =  [sig_w4+1]/sig_w4;
gan_w5  =  [sig_w5-1]/sig_w5;
gan_wm  =  [sig_wm-1]/sig_wm;
gan_we  =  [sig_we+1]/sig_we;

bet_wlab(rr,j)  =  (1+wtaulab(rr,j))*wlab0(rr,j)**(1-gan_w1)
/[(1+wtaulab(rr,j))*wlab0(rr,j)**(1-gan_w1)+(1+wtauk(rr,j)
)*wk0(rr,j)**(1-gan_w1)+(1+wtaulan(rr,j))*wlan0(rr,j)**(1-
gan_w1)+(1+wtauene(rr,j))*wene0(rr,j)**(1-gan_w1)];

bet_wlan(rr,j)  =   (1+wtaulan(rr,j))*wlan0(rr,j)**(1-gan_w1)
/[(1+wtaulab(rr,j))*wlab0(rr,j)**(1-gan_w1)+(1+wtauk(rr,j)
)*wk0(rr,j)**(1-gan_w1)+(1+wtaulan(rr,j))*wlan0(rr,j)**(1-
gan_w1)+(1+wtauene(rr,j))*wene0(rr,j)**(1-gan_w1)];

bet_wk(rr,j)  =  (1+wtauk(rr,j))*wk0(rr,j)**(1-gan_w1)
/[(1+wtaulab(rr,j))*wlab0(rr,j)**(1-gan_w1)+(1+wtauk(rr,j)
)*wk0(rr,j)**(1-gan_w1)+(1+wtaulan(rr,j))*wlan0(rr,j)**(1-
gan_w1)+(1+wtauene(rr,j))*wene0(rr,j)**(1-gan_w1)];
```

```
bet_wene(rr,j)  = (1+wtauene(rr,j))*wene0(rr,j)**(1-gan_w1)
/[(1+wtaulab(rr,j))*wlab0(rr,j)**(1-gan_w1)+(1+wtauk(rr,j)
)*wk0(rr,j)**(1-gan_w1)+(1+wtaulan(rr,j))*wlan0(rr,j)**(1-
gan_w1)+(1+wtauene(rr,j))*wene0(rr,j)**(1-gan_w1)];

alp_wy(rr,j)  =  wy0(rr,j)/{[bet_wlab(rr,j)*[wlab0(rr,j)
**gan_w1]+bet_wk(rr,j)*[wk0(rr,j)**gan_w1]+bet_wlan(rr,j)*
[wlan0(rr,j)**gan_w1]+bet_wene(rr,j)*[wene0(rr,j)**gan_w1]
]**(1/gan_w1)};

bet_wxh(rr,i)  =  (1+wtauh(rr,i))*wxh0(rr,i)/(wflab(rr)
+wflan(rr)+wfk(rr)+wfene(rr)-wtd0(rr)-wsp0(rr));

bet_wxg(rr,i)  = wxg0(rr,i)/[wtd0(rr)+sum(j,wtz0(rr,j))
+sum(h,wth0(rr,h))+sum((h,j),wtx0(rr,h,j))+sum(j,wtlab0(rr
,j))+sum(j,wtk0(rr,j))+sum(j,wtlan0(rr,j))+sum(j,wtene0(rr
,j))+sum((ss,h),wte0(rr,ss,h))+sum(h,wtej0(rr,h))+sum((ss,
h),wtm0(rr,ss,h))+sum(h,wtmj0(rr,h))-wsg0(rr)];

bet_wxi(rr,i)  = wxi0(rr,i)/(wsp0(rr)+wsg0(rr)+wsf(rr));

*ocn
bet_wee("ocn","jpn",i)  =  [(1-wtaue("ocn","jpn",i))*wee0
("ocn","jpn",i)**(1-gan_we)]
/[(1-wtaue("ocn","jpn",i))*wee0("ocn","jpn",i)**(1-gan_we)
+(1-wtaue("ocn","gca",i))*wee0("ocn","gca",i)**(1-gan_we)+
(1-wtaue("ocn","roa",i))*wee0("ocn","roa",i)**(1-gan_we)
+(1-wtaue("ocn","usa",i))*wee0("ocn","usa",i)**(1-gan_we)
+(1-wtaue("ocn","eu27",i))*wee0("ocn","eu27",i)**(1-gan_we
)+(1-wtaue("ocn","rus",i))*wee0("ocn","rus",i)**(1-gan_we)
+(1-wtaue("ocn","ind",i))*wee0("ocn","ind",i)**(1-gan_we)
+(1-wtaue("ocn","row",i))*wee0("ocn","row",i)**(1-gan_we)+
(1-wtauej("ocn",i))*ej0("ocn",i)**(1-gan_we)];
```

```
bet_wee("ocn","gca",i) = [(1-wtaue("ocn","gca",i))*wee0
("ocn","gca",i)**(1-gan_we)]/[(1-wtaue("ocn","jpn",i))*wee
0("ocn","jpn",i)**(1-gan_we)+(1-wtaue("ocn","gca",i))*wee0
("ocn","gca",i)**(1-gan_we)+(1-wtaue("ocn","roa",i))*wee0(
"ocn","roa",i)**(1-gan_we)+(1-wtaue("ocn","usa",i))*wee0("
ocn","usa",i)**(1-gan_we)+(1-wtaue("ocn","eu27",i))*wee0("
ocn","eu27",i)**(1-gan_we)+(1-wtaue("ocn","rus",i))*wee0("
ocn","rus",i)**(1-gan_we)+(1-wtaue("ocn","ind",i))*wee0("o
cn","ind",i)**(1-gan_we)+(1-wtaue("ocn","row",i))*wee0("oc
n","row",i)**(1-gan_we)+(1-wtauej("ocn",i))*ej0("ocn",i)**
(1-gan_we)];

bet_wee("ocn","roa",i) = [(1-wtaue("ocn","roa",i))*wee0
("ocn","roa",i)**(1-gan_we)]/[(1-wtaue("ocn","jpn",i))*wee
0("ocn","jpn",i)**(1-gan_we)+(1-wtaue("ocn","gca",i))*wee0
("ocn","gca",i)**(1-gan_we)+(1-wtaue("ocn","roa",i))*wee0(
"ocn","roa",i)**(1-gan_we)+(1-wtaue("ocn","usa",i))*wee0("
ocn","usa",i)**(1-gan_we)+(1-wtaue("ocn","eu27",i))*wee0("
ocn","eu27",i)**(1-gan_we)+(1-wtaue("ocn","rus",i))*wee0("
ocn","rus",i)**(1-gan_we)+(1-wtaue("ocn","ind",i))*wee0("o
cn","ind",i)**(1-gan_we)+(1-wtaue("ocn","row",i))*wee0("oc
n","row",i)**(1-gan_we)+(1-wtauej("ocn",i))*ej0("ocn",i)**
(1-gan_we)];

bet_wee("ocn","usa",i) = [(1-wtaue("ocn","usa",i))*wee0
("ocn","usa",i)**(1-gan_we)]/[(1-wtaue("ocn","jpn",i))*wee
0("ocn","jpn",i)**(1-gan_we)+(1-wtaue("ocn","gca",i))*wee0
("ocn","gca",i)**(1-gan_we)+(1-wtaue("ocn","roa",i))*wee0(
"ocn","roa",i)**(1-gan_we)+(1-wtaue("ocn","usa",i))*wee0("
ocn","usa",i)**(1-gan_we)+(1-wtaue("ocn","eu27",i))*wee0("
ocn","eu27",i)**(1-gan_we)+(1-wtaue("ocn","rus",i))*wee0("
ocn","rus",i)**(1-gan_we)+(1-wtaue("ocn","ind",i))*wee0("o
cn","ind",i)**(1-gan_we)+(1-wtaue("ocn","row",i))*wee0("oc
```

n","row",i)**(1-gan_we)+(1-wtauej("ocn",i))*ej0("ocn",i)**
(1-gan_we)];

bet_wee("ocn","eu27",i) = [(1-wtaue("ocn","eu27",i))*wee0
("ocn","eu27",i)**(1-gan_we)]/[(1-wtaue("ocn","jpn",i))*we
e0("ocn","jpn",i)**(1-gan_we)+(1-wtaue("ocn","gca",i))*wee
0("ocn","gca",i)**(1-gan_we)+(1-wtaue("ocn","roa",i))*wee0
("ocn","roa",i)**(1-gan_we)+(1-wtaue("ocn","usa",i))*wee0(
"ocn","usa",i)**(1-gan_we)+(1-wtaue("ocn","eu27",i))*wee0(
"ocn","eu27",i)**(1-gan_we)+(1-wtaue("ocn","rus",i))*wee0(
"ocn","rus",i)**(1-gan_we)+(1-wtaue("ocn","ind",i))*wee0("
ocn","ind",i)**(1-gan_we)+(1-wtaue("ocn","row",i))*wee0("o
cn","row",i)**(1-gan_we)+(1-wtauej("ocn",i))*ej0("ocn",i)*
*(1-gan_we)];

bet_wee("ocn","rus",i) = [(1-wtaue("ocn","rus",i))*wee0
("ocn","rus",i)**(1-gan_we)]/[(1-wtaue("ocn","jpn",i))*wee
0("ocn","jpn",i)**(1-gan_we)+(1-wtaue("ocn","gca",i))*wee0
("ocn","gca",i)**(1-gan_we)+(1-wtaue("ocn","roa",i))*wee0(
"ocn","roa",i)**(1-gan_we)+(1-wtaue("ocn","usa",i))*wee0("
ocn","usa",i)**(1-gan_we)+(1-wtaue("ocn","eu27",i))*wee0("
ocn","eu27",i)**(1-gan_we)+(1-wtaue("ocn","rus",i))*wee0("
ocn","rus",i)**(1-gan_we)+(1-wtaue("ocn","ind",i))*wee0("o
cn","ind",i)**(1-gan_we)
+(1-wtaue("ocn","row",i))*wee0("ocn","row",i)**(1-gan_we)+
(1-wtauej("ocn",i))*ej0("ocn",i)**(1-gan_we)];

bet_wee("ocn","ind",i) = [(1-wtaue("ocn","ind",i))*wee0
("ocn","ind",i)**(1-gan_we)]/[(1-wtaue("ocn","jpn",i))*wee
0("ocn","jpn",i)**(1-gan_we)+(1-wtaue("ocn","gca",i))*wee0
("ocn","gca",i)**(1-gan_we)+(1-wtaue("ocn","roa",i))*wee0(
"ocn","roa",i)**(1-gan_we)+(1-wtaue("ocn","usa",i))*wee0("
ocn","usa",i)**(1-gan_we)+(1-wtaue("ocn","eu27",i))*wee0("
ocn","eu27",i)**(1-gan_we)+(1-wtaue("ocn","rus",i))*wee0("

ocn","rus",i)**(1-gan_we)+(1-wtaue("ocn","ind",i))*wee0("o
cn","ind",i)**(1-gan_we)+(1-wtaue("ocn","row",i))*wee0("oc
n","row",i)**(1-gan_we)+(1-wtauej("ocn",i))*ej0("ocn",i)**
(1-gan_we)];

bet_wee("ocn","row",i) = [(1-wtaue("ocn","row",i))*wee0
("ocn","row",i)**(1-gan_we)]/[(1-wtaue("ocn","jpn",i))*wee
0("ocn","jpn",i)**(1-gan_we)+(1-wtaue("ocn","gca",i))*wee0
("ocn","gca",i)**(1-gan_we)+(1-wtaue("ocn","roa",i))*wee0(
"ocn","roa",i)**(1-gan_we)+(1-wtaue("ocn","usa",i))*wee0("
ocn","usa",i)**(1-gan_we)+(1-wtaue("ocn","eu27",i))*wee0("
ocn","eu27",i)**(1-gan_we)+(1-wtaue("ocn","rus",i))*wee0("
ocn","rus",i)**(1-gan_we)+(1-wtaue("ocn","ind",i))*wee0("o
cn","ind",i)**(1-gan_we)+(1-wtaue("ocn","row",i))*wee0("oc
n","row",i)**(1-gan_we)+(1-wtauej("ocn",i))*ej0("ocn",i)**
(1-gan_we)];

*jpn
bet_wee("jpn","ocn",i) = [(1-wtaue("jpn","ocn",i))*wee0
("jpn","ocn",i)**(1-gan_we)]/[(1-wtaue("jpn","ocn",i))*wee
0("jpn","ocn",i)**(1-gan_we)+(1-wtaue("jpn","gca",i))*wee0
("jpn","gca",i)**(1-gan_we)+(1-wtaue("jpn","roa",i))*wee0(
"jpn","roa",i)**(1-gan_we)+(1-wtaue("jpn","usa",i))*wee0("
jpn","usa",i)**(1-gan_we)+(1-wtaue("jpn","eu27",i))*wee0("
jpn","eu27",i)**(1-gan_we)+(1-wtaue("jpn","rus",i))*wee0("
jpn","rus",i)**(1-gan_we)+(1-wtaue("jpn","ind",i))*wee0("j
pn","ind",i)**(1-gan_we)+(1-wtaue("jpn","row",i))*wee0("jp
n","row",i)**(1-gan_we)+(1-wtauej("jpn",i))*ej0("jpn",i)**
(1-gan_we)];

bet_wee("jpn","gca",i) = [(1-wtaue("jpn","gca",i))*wee0
("jpn","gca",i)**(1-gan_we)]/[(1-wtaue("jpn","ocn",i))*wee
0("jpn","ocn",i)**(1-gan_we)+(1-wtaue("jpn","gca",i))*wee0
("jpn","gca",i)**(1-gan_we)+(1-wtaue("jpn","roa",i))*wee0(

"jpn","roa",i)**(1-gan_we)+(1-wtaue("jpn","usa",i))*wee0("jpn","usa",i)**(1-gan_we)+(1-wtaue("jpn","eu27",i))*wee0("jpn","eu27",i)**(1-gan_we)+(1-wtaue("jpn","rus",i))*wee0("jpn","rus",i)**(1-gan_we)+(1-wtaue("jpn","ind",i))*wee0("jpn","ind",i)**(1-gan_we)+(1-wtaue("jpn","row",i))*wee0("jpn","row",i)**(1-gan_we)+(1-wtauej("jpn",i))*ej0("jpn",i)**(1-gan_we)];

bet_wee("jpn","roa",i) = [(1-wtaue("jpn","roa",i))*wee0("jpn","roa",i)**(1-gan_we)]/[(1-wtaue("jpn","ocn",i))*wee0("jpn","ocn",i)**(1-gan_we)+(1-wtaue("jpn","gca",i))*wee0("jpn","gca",i)**(1-gan_we)+(1-wtaue("jpn","roa",i))*wee0("jpn","roa",i)**(1-gan_we)+(1-wtaue("jpn","usa",i))*wee0("jpn","usa",i)**(1-gan_we)+(1-wtaue("jpn","eu27",i))*wee0("jpn","eu27",i)**(1-gan_we)+(1-wtaue("jpn","rus",i))*wee0("jpn","rus",i)**(1-gan_we)+(1-wtaue("jpn","ind",i))*wee0("jpn","ind",i)**(1-gan_we)+(1-wtaue("jpn","row",i))*wee0("jpn","row",i)**(1-gan_we)+(1-wtauej("jpn",i))*ej0("jpn",i)**(1-gan_we)];

bet_wee("jpn","usa",i) = [(1-wtaue("jpn","usa",i))*wee0("jpn","usa",i)**(1-gan_we)]/[(1-wtaue("jpn","ocn",i))*wee0("jpn","ocn",i)**(1-gan_we)+(1-wtaue("jpn","gca",i))*wee0("jpn","gca",i)**(1-gan_we)+(1-wtaue("jpn","roa",i))*wee0("jpn","roa",i)**(1-gan_we)+(1-wtaue("jpn","usa",i))*wee0("jpn","usa",i)**(1-gan_we)+(1-wtaue("jpn","eu27",i))*wee0("jpn","eu27",i)**(1-gan_we)+(1-wtaue("jpn","rus",i))*wee0("jpn","rus",i)**(1-gan_we)+(1-wtaue("jpn","ind",i))*wee0("jpn","ind",i)**(1-gan_we)+(1-wtaue("jpn","row",i))*wee0("jpn","row",i)**(1-gan_we)+(1-wtauej("jpn",i))*ej0("jpn",i)**(1-gan_we)];

bet_wee("jpn","eu27",i) = [(1-wtaue("jpn","eu27",i))*wee0

("jpn","eu27",i)**(1-gan_we)]/[(1-wtaue("jpn","ocn",i))*we
e0("jpn","ocn",i)**(1-gan_we)+(1-wtaue("jpn","gca",i))*wee
0("jpn","gca",i)**(1-gan_we)+(1-wtaue("jpn","roa",i))*wee0
("jpn","roa",i)**(1-gan_we)+(1-wtaue("jpn","usa",i))*wee0(
"jpn","usa",i)**(1-gan_we)+(1-wtaue("jpn","eu27",i))*wee0(
"jpn","eu27",i)**(1-gan_we)+(1-wtaue("jpn","rus",i))*wee0(
"jpn","rus",i)**(1-gan_we)+(1-wtaue("jpn","ind",i))*wee0("
jpn","ind",i)**(1-gan_we)+(1-wtaue("jpn","row",i))*wee0("j
pn","row",i)**(1-gan_we)+(1-wtauej("jpn",i))*ej0("jpn",i)*
*(1-gan_we)];

bet_wee("jpn","rus",i) = [(1-wtaue("jpn","rus",i))*wee0
("jpn","rus",i)**(1-gan_we)]/[(1-wtaue("jpn","ocn",i))*wee
0("jpn","ocn",i)**(1-gan_we)+(1-wtaue("jpn","gca",i))*wee0
("jpn","gca",i)**(1-gan_we)+(1-wtaue("jpn","roa",i))*wee0(
"jpn","roa",i)**(1-gan_we)+(1-wtaue("jpn","usa",i))*wee0("
jpn","usa",i)**(1-gan_we)+(1-wtaue("jpn","eu27",i))*wee0("
jpn","eu27",i)**(1-gan_we)+(1-wtaue("jpn","rus",i))*wee0("
jpn","rus",i)**(1-gan_we)+(1-wtaue("jpn","ind",i))*wee0("j
pn","ind",i)**(1-gan_we)+(1-wtaue("jpn","row",i))*wee0("jp
n","row",i)**(1-gan_we)+(1-wtauej("jpn",i))*ej0("jpn",i)**
(1-gan_we)];

bet_wee("jpn","ind",i) = [(1-wtaue("jpn","ind",i))*wee0
("jpn","ind",i)**(1-gan_we)]/[(1-wtaue("jpn","ocn",i))*wee
0("jpn","ocn",i)**(1-gan_we)+(1-wtaue("jpn","gca",i))*wee0
("jpn","gca",i)**(1-gan_we)+(1-wtaue("jpn","roa",i))*wee0(
"jpn","roa",i)**(1-gan_we)+(1-wtaue("jpn","usa",i))*wee0("
jpn","usa",i)**(1-gan_we)+(1-wtaue("jpn","eu27",i))*wee0("
jpn","eu27",i)**(1-gan_we)+(1-wtaue("jpn","rus",i))*wee0("
jpn","rus",i)**(1-gan_we)+(1-wtaue("jpn","ind",i))*wee0("j
pn","ind",i)**(1-gan_we)+(1-wtaue("jpn","row",i))*wee0("jp
n","row",i)**(1-gan_we)+(1-wtauej("jpn",i))*ej0("jpn",i)**
(1-gan_we)];

```
bet_wee("jpn","row",i)   =   [(1-wtaue("jpn","row",i))*wee0
("jpn","row",i)**(1-gan_we)]/[(1-wtaue("jpn","ocn",i))*wee
0("jpn","ocn",i)**(1-gan_we)+(1-wtaue("jpn","gca",i))*wee0
("jpn","gca",i)**(1-gan_we)+(1-wtaue("jpn","roa",i))*wee0(
"jpn","roa",i)**(1-gan_we)+(1-wtaue("jpn","usa",i))*wee0("
jpn","usa",i)**(1-gan_we)+(1-wtaue("jpn","eu27",i))*wee0("
jpn","eu27",i)**(1-gan_we)+(1-wtaue("jpn","rus",i))*wee0("
jpn","rus",i)**(1-gan_we)+(1-wtaue("jpn","ind",i))*wee0("j
pn","ind",i)**(1-gan_we)+(1-wtaue("jpn","row",i))*wee0("jp
n","row",i)**(1-gan_we)+(1-wtauej("jpn",i))*ej0("jpn",i)**
(1-gan_we)];

*gca
bet_wee("gca","ocn",i)   =   [(1-wtaue("gca","ocn",i))*wee0
("gca","ocn",i)**(1-gan_we)]/[(1-wtaue("gca","ocn",i))*wee
0("gca","ocn",i)**(1-gan_we)+(1-wtaue("gca","jpn",i))*wee0
("gca","jpn",i)**(1-gan_we)+(1-wtaue("gca","roa",i))*wee0(
"gca","roa",i)**(1-gan_we)+(1-wtaue("gca","usa",i))*wee0("
gca","usa",i)**(1-gan_we)+(1-wtaue("gca","eu27",i))*wee0("
gca","eu27",i)**(1-gan_we)+(1-wtaue("gca","rus",i))*wee0("
gca","rus",i)**(1-gan_we)+(1-wtaue("gca","ind",i))*wee0("g
ca","ind",i)**(1-gan_we)+(1-wtaue("gca","row",i))*wee0("gc
a","row",i)**(1-gan_we)+(1-wtauej("gca",i))*ej0("gca",i)**
(1-gan_we)];

bet_wee("gca","jpn",i)   =   [(1-wtaue("gca","jpn",i))*wee0
("gca","jpn",i)**(1-gan_we)]/[(1-wtaue("gca","ocn",i))*wee
0("gca","ocn",i)**(1-gan_we)+(1-wtaue("gca","jpn",i))*wee0
("gca","jpn",i)**(1-gan_we)+(1-wtaue("gca","roa",i))*wee0(
"gca","roa",i)**(1-gan_we)+(1-wtaue("gca","usa",i))*wee0("
gca","usa",i)**(1-gan_we)+(1-wtaue("gca","eu27",i))*wee0("
gca","eu27",i)**(1-gan_we)+(1-wtaue("gca","rus",i))*wee0("
gca","rus",i)**(1-gan_we)+(1-wtaue("gca","ind",i))*wee0("g
```

```
ca","ind",i)**(1-gan_we)+(1-wtaue("gca","row",i))*wee0("gc
a","row",i)**(1-gan_we)+(1-wtauej("gca",i))*ej0("gca",i)**
(1-gan_we)];

bet_wee("gca","roa",i)  =  [(1-wtaue("gca","roa",i))*wee0
("gca","roa",i)**(1-gan_we)]/[(1-wtaue("gca","ocn",i))*wee
0("gca","ocn",i)**(1-gan_we)+(1-wtaue("gca","jpn",i))*wee0
("gca","jpn",i)**(1-gan_we)+(1-wtaue("gca","roa",i))*wee0(
"gca","roa",i)**(1-gan_we)+(1-wtaue("gca","usa",i))*wee0("
gca","usa",i)**(1-gan_we)+(1-wtaue("gca","eu27",i))*wee0("
gca","eu27",i)**(1-gan_we)+(1-wtaue("gca","rus",i))*wee0("
gca","rus",i)**(1-gan_we)+(1-wtaue("gca","ind",i))*wee0("g
ca","ind",i)**(1-gan_we)+(1-wtaue("gca","row",i))*wee0("gc
a","row",i)**(1-gan_we)+(1-wtauej("gca",i))*ej0("gca",i)**
(1-gan_we)];

bet_wee("gca","usa",i)  =  [(1-wtaue("gca","usa",i))*wee0
("gca","usa",i)**(1-gan_we)]/[(1-wtaue("gca","ocn",i))*wee
0("gca","ocn",i)**(1-gan_we)+(1-wtaue("gca","jpn",i))*wee0
("gca","jpn",i)**(1-gan_we)+(1-wtaue("gca","roa",i))*wee0(
"gca","roa",i)**(1-gan_we)+(1-wtaue("gca","usa",i))*wee0("
gca","usa",i)**(1-gan_we)+(1-wtaue("gca","eu27",i))*wee0("
gca","eu27",i)**(1-gan_we)+(1-wtaue("gca","rus",i))*wee0("
gca","rus",i)**(1-gan_we)+(1-wtaue("gca","ind",i))*wee0("g
ca","ind",i)**(1-gan_we)+(1-wtaue("gca","row",i))*wee0("gc
a","row",i)**(1-gan_we)+(1-wtauej("gca",i))*ej0("gca",i)**
(1-gan_we)];

bet_wee("gca","eu27",i)  =  [(1-wtaue("gca","eu27",i))*wee0
("gca","eu27",i)**(1-gan_we)]/[(1-wtaue("gca","ocn",i))*we
e0("gca","ocn",i)**(1-gan_we)+(1-wtaue("gca","jpn",i))*wee
0("gca","jpn",i)**(1-gan_we)+(1-wtaue("gca","roa",i))*wee0
("gca","roa",i)**(1-gan_we)+(1-wtaue("gca","usa",i))*wee0(
"gca","usa",i)**(1-gan_we)+(1-wtaue("gca","eu27",i))*wee0(
```

"gca","eu27",i)**(1-gan_we)+(1-wtaue("gca","rus",i))*wee0(
"gca","rus",i)**(1-gan_we)+(1-wtaue("gca","ind",i))*wee0("
gca","ind",i)**(1-gan_we)+(1-wtaue("gca","row",i))*wee0("g
ca","row",i)**(1-gan_we)+(1-wtauej("gca",i))*ej0("gca",i)*
*(1-gan_we)];

bet_wee("gca","rus",i) = [(1-wtaue("gca","rus",i))*wee0
("gca","rus",i)**(1-gan_we)]/[(1-wtaue("gca","ocn",i))*wee
0("gca","ocn",i)**(1-gan_we)+(1-wtaue("gca","jpn",i))*wee0
("gca","jpn",i)**(1-gan_we)+(1-wtaue("gca","roa",i))*wee0(
"gca","roa",i)**(1-gan_we)+(1-wtaue("gca","usa",i))*wee0("
gca","usa",i)**(1-gan_we)+(1-wtaue("gca","eu27",i))*wee0("
gca","eu27",i)**(1-gan_we)+(1-wtaue("gca","rus",i))*wee0("
gca","rus",i)**(1-gan_we)+(1-wtaue("gca","ind",i))*wee0("g
ca","ind",i)**(1-gan_we)+(1-wtaue("gca","row",i))*wee0("gc
a","row",i)**(1-gan_we)+(1-wtauej("gca",i))*ej0("gca",i)**
(1-gan_we)];

bet_wee("gca","ind",i) = [(1-wtaue("gca","ind",i))*wee0
("gca","ind",i)**(1-gan_we)]/[(1-wtaue("gca","ocn",i))*wee
0("gca","ocn",i)**(1-gan_we)+(1-wtaue("gca","jpn",i))*wee0
("gca","jpn",i)**(1-gan_we)+(1-wtaue("gca","roa",i))*wee0(
"gca","roa",i)**(1-gan_we)+(1-wtaue("gca","usa",i))*wee0("
gca","usa",i)**(1-gan_we)+(1-wtaue("gca","eu27",i))*wee0("
gca","eu27",i)**(1-gan_we)+(1-wtaue("gca","rus",i))*wee0("
gca","rus",i)**(1-gan_we)+(1-wtaue("gca","ind",i))*wee0("g
ca","ind",i)**(1-gan_we)+(1-wtaue("gca","row",i))*wee0("gc
a","row",i)**(1-gan_we)+(1-wtauej("gca",i))*ej0("gca",i)**
(1-gan_we)];

bet_wee("gca","row",i) = [(1-wtaue("gca","row",i))*wee0
("gca","row",i)**(1-gan_we)]/[(1-wtaue("gca","ocn",i))*wee
0("gca","ocn",i)**(1-gan_we)+(1-wtaue("gca","jpn",i))*wee0
("gca","jpn",i)**(1-gan_we)+(1-wtaue("gca","roa",i))*wee0(

"gca","roa",i)**(1-gan_we)+(1-wtaue("gca","usa",i))*wee0("
gca","usa",i)**(1-gan_we)+(1-wtaue("gca","eu27",i))*wee0("
gca","eu27",i)**(1-gan_we)+(1-wtaue("gca","rus",i))*wee0("
gca","rus",i)**(1-gan_we)+(1-wtaue("gca","ind",i))*wee0("g
ca","ind",i)**(1-gan_we)+(1-wtaue("gca","row",i))*wee0("gc
a","row",i)**(1-gan_we)+(1-wtauej("gca",i))*ej0("gca",i)**
(1-gan_we)];

*roa
bet_wee("roa","ocn",i) = [(1-wtaue("roa","ocn",i))*wee0
("roa","ocn",i)**(1-gan_we)]/[(1-wtaue("roa","ocn",i))*wee
0("roa","ocn",i)**(1-gan_we)+(1-wtaue("roa","jpn",i))*wee
0("roa","jpn",i)**(1-gan_we)+(1-wtaue("roa","gca",i))*wee
0("roa","gca",i)**(1-gan_we)+(1-wtaue("roa","usa",i))*wee
0("roa","usa",i)**(1-gan_we)+(1-wtaue("roa","eu27",i))*we
e0("roa","eu27",i)**(1-gan_we)+(1-wtaue("roa","rus",i))*w
ee0("roa","rus",i)**(1-gan_we)+(1-wtaue("roa","ind",i))*w
ee0("roa","ind",i)**(1-gan_we)+(1-wtaue("roa","row",i))*w
ee0("roa","row",i)**(1-gan_we)+(1-wtauej("roa",i))*ej0("r
oa",i)**(1-gan_we)];

bet_wee("roa","jpn",i) = [(1-wtaue("roa","jpn",i))*wee0
("roa","jpn",i)**(1-gan_we)]/[(1-wtaue("roa","ocn",i))*wee
0("roa","ocn",i)**(1-gan_we)+(1-wtaue("roa","jpn",i))*wee0
("roa","jpn",i)**(1-gan_we)+(1-wtaue("roa","gca",i))*wee0(
"roa","gca",i)**(1-gan_we)+(1-wtaue("roa","usa",i))*wee0("
roa","usa",i)**(1-gan_we)+(1-wtaue("roa","eu27",i))*wee0("
roa","eu27",i)**(1-gan_we)+(1-wtaue("roa","rus",i))*wee0("
roa","rus",i)**(1-gan_we)+(1-wtaue("roa","ind",i))*wee0("r
oa","ind",i)**(1-gan_we)+(1-wtaue("roa","row",i))*wee0("ro
a","row",i)**(1-gan_we)+(1-wtauej("roa",i))*ej0("roa",i)**
(1-gan_we)];

bet_wee("roa","gca",i) = [(1-wtaue("roa","gca",i))*wee0

("roa","gca",i)**(1-gan_we)]/[(1-wtaue("roa","ocn",i))*wee
0("roa","ocn",i)**(1-gan_we)+(1-wtaue("roa","jpn",i))*wee0
("roa","jpn",i)**(1-gan_we)+(1-wtaue("roa","gca",i))*wee0(
"roa","gca",i)**(1-gan_we)+(1-wtaue("roa","usa",i))*wee0("
roa","usa",i)**(1-gan_we)+(1-wtaue("roa","eu27",i))*wee0("
roa","eu27",i)**(1-gan_we)+(1-wtaue("roa","rus",i))*wee0("
roa","rus",i)**(1-gan_we)+(1-wtaue("roa","ind",i))*wee0("r
oa","ind",i)**(1-gan_we)+(1-wtaue("roa","row",i))*wee0("ro
a","row",i)**(1-gan_we)+(1-wtauej("roa",i))*ej0("roa",i)**
(1-gan_we)];

bet_wee("roa","usa",i) = [(1-wtaue("roa","usa",i))*wee0
("roa","usa",i)**(1-gan_we)]/[(1-wtaue("roa","ocn",i))*wee
0("roa","ocn",i)**(1-gan_we)+(1-wtaue("roa","jpn",i))*wee
0("roa","jpn",i)**(1-gan_we)+(1-wtaue("roa","gca",i))*wee
0("roa","gca",i)**(1-gan_we)+(1-wtaue("roa","usa",i))*wee
0("roa","usa",i)**(1-gan_we)+(1-wtaue("roa","eu27",i))*we
e0("roa","eu27",i)**(1-gan_we)+(1-wtaue("roa","rus",i))*w
ee0("roa","rus",i)**(1-gan_we)+(1-wtaue("roa","ind",i))*w
ee0("roa","ind",i)**(1-gan_we)+(1-wtaue("roa","row",i))*w
ee0("roa","row",i)**(1-gan_we)+(1-wtauej("roa",i))*ej0("r
oa",i)**(1-gan_we)];

bet_wee("roa","eu27",i) = [(1-wtaue("roa","eu27",i))*wee0
("roa","eu27",i)**(1-gan_we)]/[(1-wtaue("roa","ocn",i))*we
e0("roa","ocn",i)**(1-gan_we)+(1-wtaue("roa","jpn",i))*wee
0("roa","jpn",i)**(1-gan_we)+(1-wtaue("roa","gca",i))*wee0
("roa","gca",i)**(1-gan_we)+(1-wtaue("roa","usa",i))*wee0(
"roa","usa",i)**(1-gan_we)+(1-wtaue("roa","eu27",i))*wee0(
"roa","eu27",i)**(1-gan_we)+(1-wtaue("roa","rus",i))*wee0(
"roa","rus",i)**(1-gan_we)+(1-wtaue("roa","ind",i))*wee0("
roa","ind",i)**(1-gan_we)+(1-wtaue("roa","row",i))*wee0("r
oa","row",i)**(1-gan_we)+(1-wtauej("roa",i))*ej0("roa",i)*
*(1-gan_we)];

bet_wee("roa","rus",i)   =   [(1-wtaue("roa","rus",i))*wee0
("roa","rus",i)**(1-gan_we)]/[(1-wtaue("roa","ocn",i))*wee
0("roa","ocn",i)**(1-gan_we)+(1-wtaue("roa","jpn",i))*wee0
("roa","jpn",i)**(1-gan_we)+(1-wtaue("roa","gca",i))*wee0(
"roa","gca",i)**(1-gan_we)+(1-wtaue("roa","usa",i))*wee0("
roa","usa",i)**(1-gan_we)+(1-wtaue("roa","eu27",i))*wee0("
roa","eu27",i)**(1-gan_we)+(1-wtaue("roa","rus",i))*wee0("
roa","rus",i)**(1-gan_we)+(1-wtaue("roa","ind",i))*wee0("r
oa","ind",i)**(1-gan_we)+(1-wtaue("roa","row",i))*wee0("ro
a","row",i)**(1-gan_we)+(1-wtauej("roa",i))*ej0("roa",i)**
(1-gan_we)];

bet_wee("roa","ind",i)   =   [(1-wtaue("roa","ind",i))*wee0
("roa","ind",i)**(1-gan_we)]/[(1-wtaue("roa","ocn",i))*wee
0("roa","ocn",i)**(1-gan_we)+(1-wtaue("roa","jpn",i))*wee0
("roa","jpn",i)**(1-gan_we)+(1-wtaue("roa","gca",i))*wee0(
"roa","gca",i)**(1-gan_we)+(1-wtaue("roa","usa",i))*wee0("
roa","usa",i)**(1-gan_we)+(1-wtaue("roa","eu27",i))*wee0("
roa","eu27",i)**(1-gan_we)+(1-wtaue("roa","rus",i))*wee0("
roa","rus",i)**(1-gan_we)+(1-wtaue("roa","ind",i))*wee0("r
oa","ind",i)**(1-gan_we)+(1-wtaue("roa","row",i))*wee0("ro
a","row",i)**(1-gan_we)+(1-wtauej("roa",i))*ej0("roa",i)**
(1-gan_we)];

bet_wee("roa","row",i)   =   [(1-wtaue("roa","row",i))*wee0
("roa","row",i)**(1-gan_we)]/[(1-wtaue("roa","ocn",i))*wee
0("roa","ocn",i)**(1-gan_we)+(1-wtaue("roa","jpn",i))*wee0
("roa","jpn",i)**(1-gan_we)+(1-wtaue("roa","gca",i))*wee0(
"roa","gca",i)**(1-gan_we)+(1-wtaue("roa","usa",i))*wee0("
roa","usa",i)**(1-gan_we)+(1-wtaue("roa","eu27",i))*wee0("
roa","eu27",i)**(1-gan_we)+(1-wtaue("roa","rus",i))*wee0("
roa","rus",i)**(1-gan_we)+(1-wtaue("roa","ind",i))*wee0("r
oa","ind",i)**(1-gan_we)+(1-wtaue("roa","row",i))*wee0("ro

a","row",i)**(1-gan_we)+(1-wtauej("roa",i))*ej0("roa",i)**
(1-gan_we)];

*usa
bet_wee("usa","ocn",i)   =   [(1-wtaue("usa","ocn",i))*wee0
("usa","ocn",i)**(1-gan_we)]/[(1-wtaue("usa","ocn",i))*wee
0("usa","ocn",i)**(1-gan_we)+(1-wtaue("usa","jpn",i))*wee0
("usa","jpn",i)**(1-gan_we)+(1-wtaue("usa","gca",i))*wee0(
"usa","gca",i)**(1-gan_we)+(1-wtaue("usa","roa",i))*wee0("
usa","roa",i)**(1-gan_we)+(1-wtaue("usa","eu27",i))*wee0("
usa","eu27",i)**(1-gan_we)+(1-wtaue("usa","rus",i))*wee0("
usa","rus",i)**(1-gan_we)+(1-wtaue("usa","ind",i))*wee0("u
sa","ind",i)**(1-gan_we)+(1-wtaue("usa","row",i))*wee0("us
a","row",i)**(1-gan_we)+(1-wtauej("usa",i))*ej0("usa",i)**
(1-gan_we)];

bet_wee("usa","jpn",i)   =   [(1-wtaue("usa","jpn",i))*wee0
("usa","jpn",i)**(1-gan_we)]/[(1-wtaue("usa","ocn",i))*wee
0("usa","ocn",i)**(1-gan_we)+(1-wtaue("usa","jpn",i))*wee0
("usa","jpn",i)**(1-gan_we)+(1-wtaue("usa","gca",i))*wee0(
"usa","gca",i)**(1-gan_we)+(1-wtaue("usa","roa",i))*wee0("
usa","roa",i)**(1-gan_we)+(1-wtaue("usa","eu27",i))*wee0("
usa","eu27",i)**(1-gan_we)+(1-wtaue("usa","rus",i))*wee0("
usa","rus",i)**(1-gan_we)+(1-wtaue("usa","ind",i))*wee0("u
sa","ind",i)**(1-gan_we)+(1-wtaue("usa","row",i))*wee0("us
a","row",i)**(1-gan_we)+(1-wtauej("usa",i))*ej0("usa",i)**
(1-gan_we)];

bet_wee("usa","gca",i)   =   [(1-wtaue("usa","gca",i))*wee0
("usa","gca",i)**(1-gan_we)]/[(1-wtaue("usa","ocn",i))*wee
0("usa","ocn",i)**(1-gan_we)+(1-wtaue("usa","jpn",i))*wee0
("usa","jpn",i)**(1-gan_we)+(1-wtaue("usa","gca",i))*wee0(
"usa","gca",i)**(1-gan_we)+(1-wtaue("usa","roa",i))*wee0("
usa","roa",i)**(1-gan_we)+(1-wtaue("usa","eu27",i))*wee0("

```
usa","eu27",i)**(1-gan_we)+(1-wtaue("usa","rus",i))*wee0("
usa","rus",i)**(1-gan_we)+(1-wtaue("usa","ind",i))*wee0("u
sa","ind",i)**(1-gan_we)+(1-wtaue("usa","row",i))*wee0("us
a","row",i)**(1-gan_we)+(1-wtauej("usa",i))*ej0("usa",i)**
(1-gan_we)];

bet_wee("usa","roa",i)   =   [(1-wtaue("usa","roa",i))*wee0
("usa","roa",i)**(1-gan_we)]/[(1-wtaue("usa","ocn",i))*wee
0("usa","ocn",i)**(1-gan_we)+(1-wtaue("usa","jpn",i))*wee0
("usa","jpn",i)**(1-gan_we)+(1-wtaue("usa","gca",i))*wee0(
"usa","gca",i)**(1-gan_we)+(1-wtaue("usa","roa",i))*wee0("
usa","roa",i)**(1-gan_we)+(1-wtaue("usa","eu27",i))*wee0("
usa","eu27",i)**(1-gan_we)+(1-wtaue("usa","rus",i))*wee0("
usa","rus",i)**(1-gan_we)+(1-wtaue("usa","ind",i))*wee0("u
sa","ind",i)**(1-gan_we)+(1-wtaue("usa","row",i))*wee0("us
a","row",i)**(1-gan_we)+(1-wtauej("usa",i))*ej0("usa",i)**
(1-gan_we)];

bet_wee("usa","eu27",i)   =   [(1-wtaue("usa","eu27",i))*wee0
("usa","eu27",i)**(1-gan_we)]/[(1-wtaue("usa","ocn",i))*we
e0("usa","ocn",i)**(1-gan_we)+(1-wtaue("usa","jpn",i))*wee
0("usa","jpn",i)**(1-gan_we)+(1-wtaue("usa","gca",i))*wee0
("usa","gca",i)**(1-gan_we)+(1-wtaue("usa","roa",i))*wee0(
"usa","roa",i)**(1-gan_we)+(1-wtaue("usa","eu27",i))*wee0(
"usa","eu27",i)**(1-gan_we)+(1-wtaue("usa","rus",i))*wee0(
"usa","rus",i)**(1-gan_we)+(1-wtaue("usa","ind",i))*wee0("
usa","ind",i)**(1-gan_we)+(1-wtaue("usa","row",i))*wee0("u
sa","row",i)**(1-gan_we)+(1-wtauej("usa",i))*ej0("usa",i)*
*(1-gan_we)];

bet_wee("usa","rus",i)   =   [(1-wtaue("usa","rus",i))*wee0
("usa","rus",i)**(1-gan_we)]/[(1-wtaue("usa","ocn",i))*wee
0("usa","ocn",i)**(1-gan_we)+(1-wtaue("usa","jpn",i))*wee0
("usa","jpn",i)**(1-gan_we)+(1-wtaue("usa","gca",i))*wee0(
```

"usa","gca",i)**(1-gan_we)+(1-wtaue("usa","roa",i))*wee0("usa","roa",i)**(1-gan_we)+(1-wtaue("usa","eu27",i))*wee0("usa","eu27",i)**(1-gan_we)+(1-wtaue("usa","rus",i))*wee0("usa","rus",i)**(1-gan_we)+(1-wtaue("usa","ind",i))*wee0("usa","ind",i)**(1-gan_we)+(1-wtaue("usa","row",i))*wee0("usa","row",i)**(1-gan_we)+(1-wtauej("usa",i))*ej0("usa",i)**(1-gan_we)];

bet_wee("usa","ind",i) = [(1-wtaue("usa","ind",i))*wee0("usa","ind",i)**(1-gan_we)]/[(1-wtaue("usa","ocn",i))*wee0("usa","ocn",i)**(1-gan_we)+(1-wtaue("usa","jpn",i))*wee0("usa","jpn",i)**(1-gan_we)+(1-wtaue("usa","gca",i))*wee0("usa","gca",i)**(1-gan_we)+(1-wtaue("usa","roa",i))*wee0("usa","roa",i)**(1-gan_we)+(1-wtaue("usa","eu27",i))*wee0("usa","eu27",i)**(1-gan_we)+(1-wtaue("usa","rus",i))*wee0("usa","rus",i)**(1-gan_we)+(1-wtaue("usa","ind",i))*wee0("usa","ind",i)**(1-gan_we)+(1-wtaue("usa","row",i))*wee0("usa","row",i)**(1-gan_we)+(1-wtauej("usa",i))*ej0("usa",i)**(1-gan_we)];

bet_wee("usa","row",i) = [(1-wtaue("usa","row",i))*wee0("usa","row",i)**(1-gan_we)]/[(1-wtaue("usa","ocn",i))*wee0("usa","ocn",i)**(1-gan_we)+(1-wtaue("usa","jpn",i))*wee0("usa","jpn",i)**(1-gan_we)+(1-wtaue("usa","gca",i))*wee0("usa","gca",i)**(1-gan_we)+(1-wtaue("usa","roa",i))*wee0("usa","roa",i)**(1-gan_we)+(1-wtaue("usa","eu27",i))*wee0("usa","eu27",i)**(1-gan_we)+(1-wtaue("usa","rus",i))*wee0("usa","rus",i)**(1-gan_we)+(1-wtaue("usa","ind",i))*wee0("usa","ind",i)**(1-gan_we)+(1-wtaue("usa","row",i))*wee0("usa","row",i)**(1-gan_we)+(1-wtauej("usa",i))*ej0("usa",i)**(1-gan_we)];
```
```

```
*eu27
bet_wee("eu27","ocn",i)    =    [(1-wtaue("eu27","ocn",i))*wee0
("eu27","ocn",i)**(1-gan_we)]/[(1-wtaue("eu27","ocn",i))*w
ee0("eu27","ocn",i)**(1-gan_we)+(1-wtaue("eu27","jpn",i))*
wee0("eu27","jpn",i)**(1-gan_we)+(1-wtaue("eu27","gca",i))
*wee0("eu27","gca",i)**(1-gan_we)+(1-wtaue("eu27","roa",i)
)*wee0("eu27","roa",i)**(1-gan_we)+(1-wtaue("eu27","usa",i
))*wee0("eu27","usa",i)**(1-gan_we)+(1-wtaue("eu27","rus",
i))*wee0("eu27","rus",i)**(1-gan_we)+(1-wtaue("eu27","ind"
,i))*wee0("eu27","ind",i)**(1-gan_we)+(1-wtaue("eu27","row
",i))*wee0("eu27","row",i)**(1-gan_we)+(1-wtauej("eu27",i)
)*ej0("eu27",i)**(1-gan_we)];

bet_wee("eu27","jpn",i)    =    [(1-wtaue("eu27","jpn",i))*wee0
("eu27","jpn",i)**(1-gan_we)]/[(1-wtaue("eu27","ocn",i))*w
ee0("eu27","ocn",i)**(1-gan_we)+(1-wtaue("eu27","jpn",i))*
wee0("eu27","jpn",i)**(1-gan_we)+(1-wtaue("eu27","gca",i))
*wee0("eu27","gca",i)**(1-gan_we)+(1-wtaue("eu27","roa",i)
)*wee0("eu27","roa",i)**(1-gan_we)+(1-wtaue("eu27","usa",i
))*wee0("eu27","usa",i)**(1-gan_we)+(1-wtaue("eu27","rus",
i))*wee0("eu27","rus",i)**(1-gan_we)+(1-wtaue("eu27","ind"
,i))*wee0("eu27","ind",i)**(1-gan_we)+(1-wtaue("eu27","row
",i))*wee0("eu27","row",i)**(1-gan_we)+(1-wtauej("eu27",i)
)*ej0("eu27",i)**(1-gan_we)];

bet_wee("eu27","gca",i)    =    [(1-wtaue("eu27","gca",i))*wee0
("eu27","gca",i)**(1-gan_we)]/[(1-wtaue("eu27","ocn",i))*w
ee0("eu27","ocn",i)**(1-gan_we)+(1-wtaue("eu27","jpn",i))*
wee0("eu27","jpn",i)**(1-gan_we)+(1-wtaue("eu27","gca",i))
*wee0("eu27","gca",i)**(1-gan_we)+(1-wtaue("eu27","roa",i)
)*wee0("eu27","roa",i)**(1-gan_we)+(1-wtaue("eu27","usa",i
))*wee0("eu27","usa",i)**(1-gan_we)+(1-wtaue("eu27","rus",
i))*wee0("eu27","rus",i)**(1-gan_we)+(1-wtaue("eu27","ind"
,i))*wee0("eu27","ind",i)**(1-gan_we)+(1-wtaue("eu27","row
```

```
",i))*wee0("eu27","row",i)**(1-gan_we)+(1-wtauej("eu27",i)
)*ej0("eu27",i)**(1-gan_we)];

bet_wee("eu27","roa",i) = [(1-wtaue("eu27","roa",i))*wee0
("eu27","roa",i)**(1-gan_we)]/[(1-wtaue("eu27","ocn",i))*w
ee0("eu27","ocn",i)**(1-gan_we)+(1-wtaue("eu27","jpn",i))*
wee0("eu27","jpn",i)**(1-gan_we)+(1-wtaue("eu27","gca",i))
*wee0("eu27","gca",i)**(1-gan_we)+(1-wtaue("eu27","roa",i)
)*wee0("eu27","roa",i)**(1-gan_we)+(1-wtaue("eu27","usa",i
))*wee0("eu27","usa",i)**(1-gan_we)+(1-wtaue("eu27","rus",
i))*wee0("eu27","rus",i)**(1-gan_we)+(1-wtaue("eu27","ind"
,i))*wee0("eu27","ind",i)**(1-gan_we)+(1-wtaue("eu27","row
",i))*wee0("eu27","row",i)**(1-gan_we)+(1-wtauej("eu27",i)
)*ej0("eu27",i)**(1-gan_we)];

bet_wee("eu27","usa",i) = [(1-wtaue("eu27","usa",i))*wee0
("eu27","usa",i)**(1-gan_we)]/[(1-wtaue("eu27","ocn",i))*w
ee0("eu27","ocn",i)**(1-gan_we)+(1-wtaue("eu27","jpn",i))*
wee0("eu27","jpn",i)**(1-gan_we)+(1-wtaue("eu27","gca",i))
*wee0("eu27","gca",i)**(1-gan_we)+(1-wtaue("eu27","roa",i)
)*wee0("eu27","roa",i)**(1-gan_we)+(1-wtaue("eu27","usa",i
))*wee0("eu27","usa",i)**(1-gan_we)+(1-wtaue("eu27","rus",
i))*wee0("eu27","rus",i)**(1-gan_we)+(1-wtaue("eu27","ind"
,i))*wee0("eu27","ind",i)**(1-gan_we)+(1-wtaue("eu27","row
",i))*wee0("eu27","row",i)**(1-gan_we)+(1-wtauej("eu27",i)
)*ej0("eu27",i)**(1-gan_we)];

bet_wee("eu27","rus",i) = [(1-wtaue("eu27","rus",i))*wee0
("eu27","rus",i)**(1-gan_we)]/[(1-wtaue("eu27","ocn",i))*w
ee0("eu27","ocn",i)**(1-gan_we)+(1-wtaue("eu27","jpn",i))*
wee0("eu27","jpn",i)**(1-gan_we)+(1-wtaue("eu27","gca",i))
*wee0("eu27","gca",i)**(1-gan_we)+(1-wtaue("eu27","roa",i)
)*wee0("eu27","roa",i)**(1-gan_we)+(1-wtaue("eu27","usa",i
))*wee0("eu27","usa",i)**(1-gan_we)+(1-wtaue("eu27","rus",
```

i))*wee0("eu27","rus",i)**(1-gan_we)+(1-wtaue("eu27","ind"
,i))*wee0("eu27","ind",i)**(1-gan_we)+(1-wtaue("eu27","row
",i))*wee0("eu27","row",i)**(1-gan_we)+(1-wtauej("eu27",i)
)*ej0("eu27",i)**(1-gan_we)];

bet_wee("eu27","ind",i) = [(1-wtaue("eu27","ind",i))*wee0
("eu27","ind",i)**(1-gan_we)]/[(1-wtaue("eu27","ocn",i))*w
ee0("eu27","ocn",i)**(1-gan_we)+(1-wtaue("eu27","jpn",i))*
wee0("eu27","jpn",i)**(1-gan_we)+(1-wtaue("eu27","gca",i))
*wee0("eu27","gca",i)**(1-gan_we)+(1-wtaue("eu27","roa",i)
)*wee0("eu27","roa",i)**(1-gan_we)+(1-wtaue("eu27","usa",i
))*wee0("eu27","usa",i)**(1-gan_we)+(1-wtaue("eu27","rus",
i))*wee0("eu27","rus",i)**(1-gan_we)+(1-wtaue("eu27","ind"
,i))*wee0("eu27","ind",i)**(1-gan_we)+(1-wtaue("eu27","row
",i))*wee0("eu27","row",i)**(1-gan_we)+(1-wtauej("eu27",i)
)*ej0("eu27",i)**(1-gan_we)];

bet_wee("eu27","row",i) = [(1-wtaue("eu27","row",i))*wee0
("eu27","row",i)**(1-gan_we)]/[(1-wtaue("eu27","ocn",i))*w
ee0("eu27","ocn",i)**(1-gan_we)+(1-wtaue("eu27","jpn",i))*
wee0("eu27","jpn",i)**(1-gan_we)+(1-wtaue("eu27","gca",i))
*wee0("eu27","gca",i)**(1-gan_we)+(1-wtaue("eu27","roa",i)
)*wee0("eu27","roa",i)**(1-gan_we)+(1-wtaue("eu27","usa",i
))*wee0("eu27","usa",i)**(1-gan_we)+(1-wtaue("eu27","rus",
i))*wee0("eu27","rus",i)**(1-gan_we)+(1-wtaue("eu27","ind"
,i))*wee0("eu27","ind",i)**(1-gan_we)+(1-wtaue("eu27","row
",i))*wee0("eu27","row",i)**(1-gan_we)+(1-wtauej("eu27",i)
)*ej0("eu27",i)**(1-gan_we)];

*rus
bet_wee("rus","ocn",i)   =   [(1-wtaue("rus","ocn",i))*wee0
("rus","ocn",i)**(1-gan_we)]/[(1-wtaue("rus","ocn",i))*wee
0("rus","ocn",i)**(1-gan_we)+(1-wtaue("rus","jpn",i))*wee0
("rus","jpn",i)**(1-gan_we)+(1-wtaue("rus","gca",i))*wee0(

"rus","gca",i)**(1-gan_we)+(1-wtaue("rus","roa",i))*wee0("rus","roa",i)**(1-gan_we)+(1-wtaue("rus","usa",i))*wee0("rus","usa",i)**(1-gan_we)+(1-wtaue("rus","eu27",i))*wee0("rus","eu27",i)**(1-gan_we)+(1-wtaue("rus","ind",i))*wee0("rus","ind",i)**(1-gan_we)+(1-wtaue("rus","row",i))*wee0("rus","row",i)**(1-gan_we)+(1-wtauej("rus",i))*ej0("rus",i)**(1-gan_we)];

bet_wee("rus","jpn",i) = [(1-wtaue("rus","jpn",i))*wee0("rus","jpn",i)**(1-gan_we)]/[(1-wtaue("rus","ocn",i))*wee0("rus","ocn",i)**(1-gan_we)+(1-wtaue("rus","jpn",i))*wee0("rus","jpn",i)**(1-gan_we)+(1-wtaue("rus","gca",i))*wee0("rus","gca",i)**(1-gan_we)+(1-wtaue("rus","roa",i))*wee0("rus","roa",i)**(1-gan_we)+(1-wtaue("rus","usa",i))*wee0("rus","usa",i)**(1-gan_we)+(1-wtaue("rus","eu27",i))*wee0("rus","eu27",i)**(1-gan_we)+(1-wtaue("rus","ind",i))*wee0("rus","ind",i)**(1-gan_we)+(1-wtaue("rus","row",i))*wee0("rus","row",i)**(1-gan_we)+(1-wtauej("rus",i))*ej0("rus",i)**(1-gan_we)];

bet_wee("rus","gca",i) = [(1-wtaue("rus","gca",i))*wee0("rus","gca",i)**(1-gan_we)]/[(1-wtaue("rus","ocn",i))*wee0("rus","ocn",i)**(1-gan_we)+(1-wtaue("rus","jpn",i))*wee0("rus","jpn",i)**(1-gan_we)+(1-wtaue("rus","gca",i))*wee0("rus","gca",i)**(1-gan_we)+(1-wtaue("rus","roa",i))*wee0("rus","roa",i)**(1-gan_we)+(1-wtaue("rus","usa",i))*wee0("rus","usa",i)**(1-gan_we)+(1-wtaue("rus","eu27",i))*wee0("rus","eu27",i)**(1-gan_we)+(1-wtaue("rus","ind",i))*wee0("rus","ind",i)**(1-gan_we)+(1-wtaue("rus","row",i))*wee0("rus","row",i)**(1-gan_we)+(1-wtauej("rus",i))*ej0("rus",i)**(1-gan_we)];

bet_wee("rus","roa",i) = [(1-wtaue("rus","roa",i))*wee0("rus","roa",i)**(1-gan_we)]/[(1-wtaue("rus","ocn",i))*wee

```
0("rus","ocn",i)**(1-gan_we)+(1-wtaue("rus","jpn",i))*wee0
("rus","jpn",i)**(1-gan_we)+(1-wtaue("rus","gca",i))*wee0(
"rus","gca",i)**(1-gan_we)+(1-wtaue("rus","roa",i))*wee0("
rus","roa",i)**(1-gan_we)+(1-wtaue("rus","usa",i))*wee0("r
us","usa",i)**(1-gan_we)+(1-wtaue("rus","eu27",i))*wee0("r
us","eu27",i)**(1-gan_we)+(1-wtaue("rus","ind",i))*wee0("r
us","ind",i)**(1-gan_we)+(1-wtaue("rus","row",i))*wee0("ru
s","row",i)**(1-gan_we)+(1-wtauej("rus",i))*ej0("rus",i)**
(1-gan_we)];

bet_wee("rus","usa",i)   =   [(1-wtaue("rus","usa",i))*wee0
("rus","usa",i)**(1-gan_we)]/[(1-wtaue("rus","ocn",i))*wee
0("rus","ocn",i)**(1-gan_we)+(1-wtaue("rus","jpn",i))*wee0
("rus","jpn",i)**(1-gan_we)+(1-wtaue("rus","gca",i))*wee0(
"rus","gca",i)**(1-gan_we)+(1-wtaue("rus","roa",i))*wee0("
rus","roa",i)**(1-gan_we)+(1-wtaue("rus","usa",i))*wee0("r
us","usa",i)**(1-gan_we)+(1-wtaue("rus","eu27",i))*wee0("r
us","eu27",i)**(1-gan_we)+(1-wtaue("rus","ind",i))*wee0("r
us","ind",i)**(1-gan_we)+(1-wtaue("rus","row",i))*wee0("ru
s","row",i)**(1-gan_we)+(1-wtauej("rus",i))*ej0("rus",i)**
(1-gan_we)];

bet_wee("rus","eu27",i)   =   [(1-wtaue("rus","eu27",i))*wee0
("rus","eu27",i)**(1-gan_we)]/[(1-wtaue("rus","ocn",i))*we
e0("rus","ocn",i)**(1-gan_we)+(1-wtaue("rus","jpn",i))*wee
0("rus","jpn",i)**(1-gan_we)+(1-wtaue("rus","gca",i))*wee0
("rus","gca",i)**(1-gan_we)+(1-wtaue("rus","roa",i))*wee0(
"rus","roa",i)**(1-gan_we)+(1-wtaue("rus","usa",i))*wee0("
rus","usa",i)**(1-gan_we)+(1-wtaue("rus","eu27",i))*wee0("
rus","eu27",i)**(1-gan_we)+(1-wtaue("rus","ind",i))*wee0("
rus","ind",i)**(1-gan_we)+(1-wtaue("rus","row",i))*wee0("r
us","row",i)**(1-gan_we)+(1-wtauej("rus",i))*ej0("rus",i)*
*(1-gan_we)];
```

bet_wee("rus","ind",i)   =   [(1-wtaue("rus","ind",i))*wee0("rus","ind",i)**(1-gan_we)]/[(1-wtaue("rus","ocn",i))*wee0("rus","ocn",i)**(1-gan_we)+(1-wtaue("rus","jpn",i))*wee0("rus","jpn",i)**(1-gan_we)+(1-wtaue("rus","gca",i))*wee0("rus","gca",i)**(1-gan_we)+(1-wtaue("rus","roa",i))*wee0("rus","roa",i)**(1-gan_we)+(1-wtaue("rus","usa",i))*wee0("rus","usa",i)**(1-gan_we)+(1-wtaue("rus","eu27",i))*wee0("rus","eu27",i)**(1-gan_we)+(1-wtaue("rus","ind",i))*wee0("rus","ind",i)**(1-gan_we)+(1-wtaue("rus","row",i))*wee0("rus","row",i)**(1-gan_we)+(1-wtauej("rus",i))*ej0("rus",i)**(1-gan_we)];

bet_wee("rus","row",i)   =   [(1-wtaue("rus","row",i))*wee0("rus","row",i)**(1-gan_we)]/[(1-wtaue("rus","ocn",i))*wee0("rus","ocn",i)**(1-gan_we)+(1-wtaue("rus","jpn",i))*wee0("rus","jpn",i)**(1-gan_we)+(1-wtaue("rus","gca",i))*wee0("rus","gca",i)**(1-gan_we)+(1-wtaue("rus","roa",i))*wee0("rus","roa",i)**(1-gan_we)+(1-wtaue("rus","usa",i))*wee0("rus","usa",i)**(1-gan_we)+(1-wtaue("rus","eu27",i))*wee0("rus","eu27",i)**(1-gan_we)+(1-wtaue("rus","ind",i))*wee0("rus","ind",i)**(1-gan_we)+(1-wtaue("rus","row",i))*wee0("rus","row",i)**(1-gan_we)+(1-wtauej("rus",i))*ej0("rus",i)**(1-gan_we)];

*ind
bet_wee("ind","ocn",i)   =   [(1-wtaue("ind","ocn",i))*wee0("ind","ocn",i)**(1-gan_we)]/[(1-wtaue("ind","ocn",i))*wee0("ind","ocn",i)**(1-gan_we)+(1-wtaue("ind","jpn",i))*wee0("ind","jpn",i)**(1-gan_we)+(1-wtaue("ind","gca",i))*wee0("ind","gca",i)**(1-gan_we)+(1-wtaue("ind","roa",i))*wee0("ind","roa",i)**(1-gan_we)+(1-wtaue("ind","usa",i))*wee0("ind","usa",i)**(1-gan_we)+(1-wtaue("ind","eu27",i))*wee0("ind","eu27",i)**(1-gan_we)+(1-wtaue("ind","rus",i))*wee0("ind","rus",i)**(1-gan_we)+(1-wtaue("ind","row",i))*wee0("in

d","row",i)**(1-gan_we)+(1-wtauej("ind",i))*ej0("ind",i)**
(1-gan_we)];

bet_wee("ind","jpn",i)   =   [(1-wtaue("ind","jpn",i))*wee0
("ind","jpn",i)**(1-gan_we)]/[(1-wtaue("ind","ocn",i))*wee
0("ind","ocn",i)**(1-gan_we)+(1-wtaue("ind","jpn",i))*wee0
("ind","jpn",i)**(1-gan_we)+(1-wtaue("ind","gca",i))*wee0(
"ind","gca",i)**(1-gan_we)+(1-wtaue("ind","roa",i))*wee0("
ind","roa",i)**(1-gan_we)+(1-wtaue("ind","usa",i))*wee0("i
nd","usa",i)**(1-gan_we)+(1-wtaue("ind","eu27",i))*wee0("i
nd","eu27",i)**(1-gan_we)+(1-wtaue("ind","rus",i))*wee0("i
nd","rus",i)**(1-gan_we)+(1-wtaue("ind","row",i))*wee0("in
d","row",i)**(1-gan_we)+(1-wtauej("ind",i))*ej0("ind",i)**
(1-gan_we)];

bet_wee("ind","gca",i)   =   [(1-wtaue("ind","gca",i))*wee0
("ind","gca",i)**(1-gan_we)]/[(1-wtaue("ind","ocn",i))*wee
0("ind","ocn",i)**(1-gan_we)+(1-wtaue("ind","jpn",i))*wee0
("ind","jpn",i)**(1-gan_we)+(1-wtaue("ind","gca",i))*wee0(
"ind","gca",i)**(1-gan_we)+(1-wtaue("ind","roa",i))*wee0("
ind","roa",i)**(1-gan_we)+(1-wtaue("ind","usa",i))*wee0("i
nd","usa",i)**(1-gan_we)+(1-wtaue("ind","eu27",i))*wee0("i
nd","eu27",i)**(1-gan_we)+(1-wtaue("ind","rus",i))*wee0("i
nd","rus",i)**(1-gan_we)+(1-wtaue("ind","row",i))*wee0("in
d","row",i)**(1-gan_we)+(1-wtauej("ind",i))*ej0("ind",i)**
(1-gan_we)];

bet_wee("ind","roa",i)   =   [(1-wtaue("ind","roa",i))*wee0
("ind","roa",i)**(1-gan_we)]/[(1-wtaue("ind","ocn",i))*wee
0("ind","ocn",i)**(1-gan_we)+(1-wtaue("ind","jpn",i))*wee0
("ind","jpn",i)**(1-gan_we)+(1-wtaue("ind","gca",i))*wee0(
"ind","gca",i)**(1-gan_we)+(1-wtaue("ind","roa",i))*wee0("
ind","roa",i)**(1-gan_we)+(1-wtaue("ind","usa",i))*wee0("i
nd","usa",i)**(1-gan_we)+(1-wtaue("ind","eu27",i))*wee0("i

nd","eu27",i)**(1-gan_we)+(1-wtaue("ind","rus",i))*wee0("ind","rus",i)**(1-gan_we)+(1-wtaue("ind","row",i))*wee0("ind","row",i)**(1-gan_we)+(1-wtauej("ind",i))*ej0("ind",i)**(1-gan_we)];

bet_wee("ind","usa",i)   =   [(1-wtaue("ind","usa",i))*wee0("ind","usa",i)**(1-gan_we)]/[(1-wtaue("ind","ocn",i))*wee0("ind","ocn",i)**(1-gan_we)+(1-wtaue("ind","jpn",i))*wee0("ind","jpn",i)**(1-gan_we)+(1-wtaue("ind","gca",i))*wee0("ind","gca",i)**(1-gan_we)+(1-wtaue("ind","roa",i))*wee0("ind","roa",i)**(1-gan_we)+(1-wtaue("ind","usa",i))*wee0("ind","usa",i)**(1-gan_we)+(1-wtaue("ind","eu27",i))*wee0("ind","eu27",i)**(1-gan_we)+(1-wtaue("ind","rus",i))*wee0("ind","rus",i)**(1-gan_we)+(1-wtaue("ind","row",i))*wee0("ind","row",i)**(1-gan_we)+(1-wtauej("ind",i))*ej0("ind",i)**(1-gan_we)];

bet_wee("ind","eu27",i) = [(1-wtaue("ind","eu27",i))*wee0("ind","eu27",i)**(1-gan_we)]/[(1-wtaue("ind","ocn",i))*wee0("ind","ocn",i)**(1-gan_we)+(1-wtaue("ind","jpn",i))*wee0("ind","jpn",i)**(1-gan_we)+(1-wtaue("ind","gca",i))*wee0("ind","gca",i)**(1-gan_we)+(1-wtaue("ind","roa",i))*wee0("ind","roa",i)**(1-gan_we)+(1-wtaue("ind","usa",i))*wee0("ind","usa",i)**(1-gan_we)+(1-wtaue("ind","eu27",i))*wee0("ind","eu27",i)**(1-gan_we)+(1-wtaue("ind","rus",i))*wee0("ind","rus",i)**(1-gan_we)+(1-wtaue("ind","row",i))*wee0("ind","row",i)**(1-gan_we)+(1-wtauej("ind",i))*ej0("ind",i)**(1-gan_we)];

bet_wee("ind","rus",i)   =   [(1-wtaue("ind","rus",i))*wee0("ind","rus",i)**(1-gan_we)]/[(1-wtaue("ind","ocn",i))*wee0("ind","ocn",i)**(1-gan_we)+(1-wtaue("ind","jpn",i))*wee0("ind","jpn",i)**(1-gan_we)+(1-wtaue("ind","gca",i))*wee0("ind","gca",i)**(1-gan_we)+(1-wtaue("ind","roa",i))*wee0("

```
ind","roa",i)**(1-gan_we)+(1-wtaue("ind","usa",i))*wee0("i
nd","usa",i)**(1-gan_we)+(1-wtaue("ind","eu27",i))*wee0("i
nd","eu27",i)**(1-gan_we)+(1-wtaue("ind","rus",i))*wee0("i
nd","rus",i)**(1-gan_we)+(1-wtaue("ind","row",i))*wee0("in
d","row",i)**(1-gan_we)+(1-wtauej("ind",i))*ej0("ind",i)**
(1-gan_we)];

bet_wee("ind","row",i)   =   [(1-wtaue("ind","row",i))*wee0
("ind","row",i)**(1-gan_we)]/[(1-wtaue("ind","ocn",i))*wee
0("ind","ocn",i)**(1-gan_we)+(1-wtaue("ind","jpn",i))*wee0
("ind","jpn",i)**(1-gan_we)+(1-wtaue("ind","gca",i))*wee0(
"ind","gca",i)**(1-gan_we)+(1-wtaue("ind","roa",i))*wee0("
ind","roa",i)**(1-gan_we)+(1-wtaue("ind","usa",i))*wee0("i
nd","usa",i)**(1-gan_we)+(1-wtaue("ind","eu27",i))*wee0("i
nd","eu27",i)**(1-gan_we)+(1-wtaue("ind","rus",i))*wee0("i
nd","rus",i)**(1-gan_we)+(1-wtaue("ind","row",i))*wee0("in
d","row",i)**(1-gan_we)+(1-wtauej("ind",i))*ej0("ind",i)**
(1-gan_we)];

*row
bet_wee("row","ocn",i)   =   [(1-wtaue("row","ocn",i))*wee0
("row","ocn",i)**(1-gan_we)]/[(1-wtaue("row","ocn",i))*wee
0("row","ocn",i)**(1-gan_we)+(1-wtaue("row","jpn",i))*wee0
("row","jpn",i)**(1-gan_we)+(1-wtaue("row","gca",i))*wee0(
"row","gca",i)**(1-gan_we)+(1-wtaue("row","roa",i))*wee0("
row","roa",i)**(1-gan_we)+(1-wtaue("row","usa",i))*wee0("r
ow","usa",i)**(1-gan_we)+(1-wtaue("row","eu27",i))*wee0("r
ow","eu27",i)**(1-gan_we)+(1-wtaue("row","rus",i))*wee0("r
ow","rus",i)**(1-gan_we)+(1-wtaue("row","ind",i))*wee0("ro
w","ind",i)**(1-gan_we)+(1-wtauej("row",i))*ej0("row",i)**
(1-gan_we)];

bet_wee("row","jpn",i)   =   [(1-wtaue("row","jpn",i))*wee0
("row","jpn",i)**(1-gan_we)]/[(1-wtaue("row","ocn",i))*wee
```

0("row","ocn",i)**(1-gan_we)+(1-wtaue("row","jpn",i))*wee0
("row","jpn",i)**(1-gan_we)+(1-wtaue("row","gca",i))*wee0(
"row","gca",i)**(1-gan_we)+(1-wtaue("row","roa",i))*wee0("
row","roa",i)**(1-gan_we)+(1-wtaue("row","usa",i))*wee0("r
ow","usa",i)**(1-gan_we)+(1-wtaue("row","eu27",i))*wee0("r
ow","eu27",i)**(1-gan_we)+(1-wtaue("row","rus",i))*wee0("r
ow","rus",i)**(1-gan_we)+(1-wtaue("row","ind",i))*wee0("ro
w","ind",i)**(1-gan_we)+(1-wtauej("row",i))*ej0("row",i)**
(1-gan_we)];

bet_wee("row","gca",i)   =   [(1-wtaue("row","gca",i))*wee0
("row","gca",i)**(1-gan_we)]/[(1-wtaue("row","ocn",i))*wee
0("row","ocn",i)**(1-gan_we)+(1-wtaue("row","jpn",i))*wee0
("row","jpn",i)**(1-gan_we)+(1-wtaue("row","gca",i))*wee0(
"row","gca",i)**(1-gan_we)+(1-wtaue("row","roa",i))*wee0("
row","roa",i)**(1-gan_we)+(1-wtaue("row","usa",i))*wee0("r
ow","usa",i)**(1-gan_we)+(1-wtaue("row","eu27",i))*wee0("r
ow","eu27",i)**(1-gan_we)+(1-wtaue("row","rus",i))*wee0("r
ow","rus",i)**(1-gan_we)+(1-wtaue("row","ind",i))*wee0("ro
w","ind",i)**(1-gan_we)+(1-wtauej("row",i))*ej0("row",i)**
(1-gan_we)];

bet_wee("row","roa",i)   =   [(1-wtaue("row","roa",i))*wee0
("row","roa",i)**(1-gan_we)]/[(1-wtaue("row","ocn",i))*wee
0("row","ocn",i)**(1-gan_we)+(1-wtaue("row","jpn",i))*wee0
("row","jpn",i)**(1-gan_we)+(1-wtaue("row","gca",i))*wee0(
"row","gca",i)**(1-gan_we)+(1-wtaue("row","roa",i))*wee0("
row","roa",i)**(1-gan_we)+(1-wtaue("row","usa",i))*wee0("r
ow","usa",i)**(1-gan_we)+(1-wtaue("row","eu27",i))*wee0("r
ow","eu27",i)**(1-gan_we)+(1-wtaue("row","rus",i))*wee0("r
ow","rus",i)**(1-gan_we)+(1-wtaue("row","ind",i))*wee0("ro
w","ind",i)**(1-gan_we)+(1-wtauej("row",i))*ej0("row",i)**
(1-gan_we)];

```
bet_wee("row","usa",i)    =    [(1-wtaue("row","usa",i))*wee0
("row","usa",i)**(1-gan_we)]/[(1-wtaue("row","ocn",i))*wee
0("row","ocn",i)**(1-gan_we)+(1-wtaue("row","jpn",i))*wee0
("row","jpn",i)**(1-gan_we)+(1-wtaue("row","gca",i))*wee0(
"row","gca",i)**(1-gan_we)+(1-wtaue("row","roa",i))*wee0("
row","roa",i)**(1-gan_we)+(1-wtaue("row","usa",i))*wee0("r
ow","usa",i)**(1-gan_we)+(1-wtaue("row","eu27",i))*wee0("r
ow","eu27",i)**(1-gan_we)+(1-wtaue("row","rus",i))*wee0("r
ow","rus",i)**(1-gan_we)+(1-wtaue("row","ind",i))*wee0("ro
w","ind",i)**(1-gan_we)+(1-wtauej("row",i))*ej0("row",i)**
(1-gan_we)];


bet_wee("row","eu27",i)   =    [(1-wtaue("row","eu27",i))*wee0
("row","eu27",i)**(1-gan_we)]/[(1-wtaue("row","ocn",i))*we
e0("row","ocn",i)**(1-gan_we)+(1-wtaue("row","jpn",i))*wee
0("row","jpn",i)**(1-gan_we)+(1-wtaue("row","gca",i))*wee0
("row","gca",i)**(1-gan_we)+(1-wtaue("row","roa",i))*wee0(
"row","roa",i)**(1-gan_we)+(1-wtaue("row","usa",i))*wee0("
row","usa",i)**(1-gan_we)+(1-wtaue("row","eu27",i))*wee0("
row","eu27",i)**(1-gan_we)+(1-wtaue("row","rus",i))*wee0("
row","rus",i)**(1-gan_we)+(1-wtaue("row","ind",i))*wee0("r
ow","ind",i)**(1-gan_we)+(1-wtauej("row",i))*ej0("row",i)*
*(1-gan_we)];


bet_wee("row","rus",i)    =    [(1-wtaue("row","rus",i))*wee0
("row","rus",i)**(1-gan_we)]/[(1-wtaue("row","ocn",i))*wee
0("row","ocn",i)**(1-gan_we)+(1-wtaue("row","jpn",i))*wee0
("row","jpn",i)**(1-gan_we)+(1-wtaue("row","gca",i))*wee0(
"row","gca",i)**(1-gan_we)+(1-wtaue("row","roa",i))*wee0("
row","roa",i)**(1-gan_we)+(1-wtaue("row","usa",i))*wee0("r
ow","usa",i)**(1-gan_we)+(1-wtaue("row","eu27",i))*wee0("r
ow","eu27",i)**(1-gan_we)+(1-wtaue("row","rus",i))*wee0("r
ow","rus",i)**(1-gan_we)+(1-wtaue("row","ind",i))*wee0("ro
```

w","ind",i)**(1-gan_we)+(1-wtauej("row",i))*ej0("row",i)**
(1-gan_we)];

bet_wee("row","ind",i) = [(1-wtaue("row","ind",i))*wee0
("row","ind",i)**(1-gan_we)]/[(1-wtaue("row","ocn",i))*wee
0("row","ocn",i)**(1-gan_we)+(1-wtaue("row","jpn",i))*wee0
("row","jpn",i)**(1-gan_we)+(1-wtaue("row","gca",i))*wee0(
"row","gca",i)**(1-gan_we)+(1-wtaue("row","roa",i))*wee0("
row","roa",i)**(1-gan_we)+(1-wtaue("row","usa",i))*wee0("r
ow","usa",i)**(1-gan_we)+(1-wtaue("row","eu27",i))*wee0("r
ow","eu27",i)**(1-gan_we)+(1-wtaue("row","rus",i))*wee0("r
ow","rus",i)**(1-gan_we)+(1-wtaue("row","ind",i))*wee0("ro
w","ind",i)**(1-gan_we)+(1-wtauej("row",i))*ej0("row",i)**
(1-gan_we)];

*bet_ej
bet_ej("ocn",i) = z(1-wtauej("ocn",i))*ej0("ocn",i)**(1-gan_we)]
/[(1-wtaue("ocn","jpn",i))*wee0("ocn","jpn",i)**(1-gan_we)
+(1-wtaue("ocn","gca",i))*wee0("ocn","gca",i)**(1-gan_we)+
(1-wtaue("ocn","roa",i))*wee0("ocn","roa",i)**(1-gan_we)
+(1-wtaue("ocn","usa",i))*wee0("ocn","usa",i)**(1-gan_we)
+(1-wtaue("ocn","eu27",i))*wee0("ocn","eu27",i)**(1-gan_we
)+(1-wtaue("ocn","rus",i))*wee0("ocn","rus",i)**(1-gan_we)
+(1-wtaue("ocn","ind",i))*wee0("ocn","ind",i)**(1-gan_we)
+(1-wtaue("ocn","row",i))*wee0("ocn","row",i)**(1-gan_we)+
(1-wtauej("ocn",i))*ej0("ocn",i)**(1-gan_we)];

bet_ej("jpn",i) = [(1-wtauej("jpn",i))*ej0("jpn",i)**(1-gan_we)]
/[(1-wtaue("jpn","ocn",i))*wee0("jpn","ocn",i)**(1-gan_we)
+(1-wtaue("jpn","gca",i))*wee0("jpn","gca",i)**(1-gan_we)+
(1-wtaue("jpn","roa",i))*wee0("jpn","roa",i)**(1-gan_we)
+(1-wtaue("jpn","usa",i))*wee0("jpn","usa",i)**(1-gan_we)

+(1-wtaue("jpn","eu27",i))*wee0("jpn","eu27",i)**(1-gan_we
)+(1-wtaue("jpn","rus",i))*wee0("jpn","rus",i)**(1-gan_we)
+(1-wtaue("jpn","ind",i))*wee0("jpn","ind",i)**(1-gan_we)
+(1-wtaue("jpn","row",i))*wee0("jpn","row",i)**(1-gan_we)+
(1-wtauej("jpn",i))*ej0("jpn",i)**(1-gan_we)];

bet_ej("gca",i) = [(1-wtauej("gca",i))*ej0("gca",i)**(1-gan_we)]
/[(1-wtaue("gca","ocn",i))*wee0("gca","ocn",i)**(1-gan_we)
+(1-wtaue("gca","jpn",i))*wee0("gca","jpn",i)**(1-gan_we)+
(1-wtaue("gca","roa",i))*wee0("gca","roa",i)**(1-gan_we)
+(1-wtaue("gca","usa",i))*wee0("gca","usa",i)**(1-gan_we)
+(1-wtaue("gca","eu27",i))*wee0("gca","eu27",i)**(1-gan_we
)+(1-wtaue("gca","rus",i))*wee0("gca","rus",i)**(1-gan_we)
+(1-wtaue("gca","ind",i))*wee0("gca","ind",i)**(1-gan_we)
+(1-wtaue("gca","row",i))*wee0("gca","row",i)**(1-gan_we)+
(1-wtauej("gca",i))*ej0("gca",i)**(1-gan_we)];

bet_ej("roa",i) = [(1-wtauej("roa",i))*ej0("roa",i)**(1-gan_we)]
/[(1-wtaue("roa","ocn",i))*wee0("roa","ocn",i)**(1-gan_we)
+(1-wtaue("roa","jpn",i))*wee0("roa","jpn",i)**(1-gan_we)+
(1-wtaue("roa","gca",i))*wee0("roa","gca",i)**(1-gan_we)
+(1-wtaue("roa","usa",i))*wee0("roa","usa",i)**(1-gan_we)
+(1-wtaue("roa","eu27",i))*wee0("roa","eu27",i)**(1-gan_we
)+(1-wtaue("roa","rus",i))*wee0("roa","rus",i)**(1-gan_we)
+(1-wtaue("roa","ind",i))*wee0("roa","ind",i)**(1-gan_we)
+(1-wtaue("roa","row",i))*wee0("roa","row",i)**(1-gan_we)+
(1-wtauej("roa",i))*ej0("roa",i)**(1-gan_we)];

bet_ej("usa",i) = [(1-wtauej("usa",i))*ej0("usa",i)**(1-gan_we)]
/[(1-wtaue("usa","ocn",i))*wee0("usa","ocn",i)**(1-gan_we)
+(1-wtaue("usa","jpn",i))*wee0("usa","jpn",i)**(1-gan_we)+
(1-wtaue("usa","gca",i))*wee0("usa","gca",i)**(1-gan_we)
+(1-wtaue("usa","roa",i))*wee0("usa","roa",i)**(1-gan_we)

```
+(1-wtaue("usa","eu27",i))*wee0("usa","eu27",i)**(1-gan_we
)+(1-wtaue("usa","rus",i))*wee0("usa","rus",i)**(1-gan_we)
+(1-wtaue("usa","ind",i))*wee0("usa","ind",i)**(1-gan_we)
+(1-wtaue("usa","row",i))*wee0("usa","row",i)**(1-gan_we)+
(1-wtauej("usa",i))*ej0("usa",i)**(1-gan_we)];

bet_ej("eu27",i)  = [(1-wtauej("eu27",i))*ej0("eu27",i)
**(1-gan_we)]/[(1-wtaue("eu27","ocn",i))*wee0("eu27","ocn"
,i)**(1-gan_we)+(1-wtaue("eu27","jpn",i))*wee0("eu27","jpn
",i)**(1-gan_we)+(1-wtaue("eu27","gca",i))*wee0("eu27","gc
a",i)**(1-gan_we)+(1-wtaue("eu27","roa",i))*wee0("eu27","r
oa",i)**(1-gan_we)+(1-wtaue("eu27","usa",i))*wee0("eu27","
usa",i)**(1-gan_we)+(1-wtaue("eu27","rus",i))*wee0("eu27",
"rus",i)**(1-gan_we)+(1-wtaue("eu27","ind",i))*wee0("eu27"
,"ind",i)**(1-gan_we)+(1-wtaue("eu27","row",i))*wee0("eu27
","row",i)**(1-gan_we)+(1-wtauej("eu27",i))*ej0("eu27",i)*
*(1-gan_we)];

bet_ej("rus",i)  =  [(1-wtauej("rus",i))*ej0("rus",i)
**(1-gan_we)]/[(1-wtaue("rus","ocn",i))*wee0("rus","ocn",i
)**(1-gan_we)+(1-wtaue("rus","jpn",i))*wee0("rus","jpn",i)
**(1-gan_we)+(1-wtaue("rus","gca",i))*wee0("rus","gca",i)*
*(1-gan_we)+(1-wtaue("rus","roa",i))*wee0("rus","roa",i)**
(1-gan_we)+(1-wtaue("rus","usa",i))*wee0("rus","usa",i)**(
1-gan_we)+(1-wtaue("rus","eu27",i))*wee0("rus","eu27",i)**
(1-gan_we)+(1-wtaue("rus","ind",i))*wee0("rus","ind",i)**(
1-gan_we)+(1-wtaue("rus","row",i))*wee0("rus","row",i)**(1
-gan_we)+(1-wtauej("rus",i))*ej0("rus",i)**(1-gan_we)];

bet_ej("ind",i)  =  [(1-wtauej("ind",i))*ej0("ind",i)
**(1-gan_we)]/[(1-wtaue("ind","ocn",i))*wee0("ind","ocn",i
)**(1-gan_we)+(1-wtaue("ind","jpn",i))*wee0("ind","jpn",i)
**(1-gan_we)+(1-wtaue("ind","gca",i))*wee0("ind","gca",i)*
*(1-gan_we)+(1-wtaue("ind","roa",i))*wee0("ind","roa",i)**
```

```
(1-gan_we)+(1-wtaue("ind","usa",i))*wee0("ind","usa",i)**(
1-gan_we)+(1-wtaue("ind","eu27",i))*wee0("ind","eu27",i)**
(1-gan_we)+(1-wtaue("ind","rus",i))*wee0("ind","rus",i)**(
1-gan_we)+(1-wtaue("ind","row",i))*wee0("ind","row",i)**(1
-gan_we)+(1-wtauej("ind",i))*ej0("ind",i)**(1-gan_we)];

bet_ej("row",i)   =   [(1-wtauej("row",i))*ej0("row",i)
**(1-gan_we)]/[(1-wtaue("row","ocn",i))*wee0("row","ocn",i
)**(1-gan_we)+(1-wtaue("row","jpn",i))*wee0("row","jpn",i)
**(1-gan_we)+(1-wtaue("row","gca",i))*wee0("row","gca",i)*
*(1-gan_we)+(1-wtaue("row","roa",i))*wee0("row","roa",i)**
(1-gan_we)+(1-wtaue("row","usa",i))*wee0("row","usa",i)**(
1-gan_we)+(1-wtaue("row","eu27",i))*wee0("row","eu27",i)**
(1-gan_we)+(1-wtaue("row","rus",i))*wee0("row","rus",i)**(
1-gan_we)+(1-wtaue("row","ind",i))*wee0("row","ind",i)**(1
-gan_we)+(1-wtauej("row",i))*ej0("row",i)**(1-gan_we)];

bet_wee("ocn","ocn",i)    = 0;
bet_wee("jpn","jpn",i)    = 0;
bet_wee("gca","gca",i)    = 0;
bet_wee("roa","roa",i)    = 0;
bet_wee("usa","usa",i)    = 0;
bet_wee("eu27","eu27",i)  = 0;
bet_wee("rus","rus",i)    = 0;
bet_wee("ind","ind",i)    = 0;
bet_wee("row","row",i)    = 0;

alp_wee(rr,i)   =   we0(rr,i)/{[sum(ss,bet_wee(rr,ss,i)*wee0
(rr,ss,i)**gan_we)+bet_ej(rr,i)*ej0(rr,i)**gan_we]**(1/gan
_we)};

bet_we(rr,i)  =  [we0(rr,i)**(1-gan_w4)]/[we0(rr,i)**(1-gan_w4)
+wd0(rr,i)**(1-gan_w4)];
```

```
bet_wd(rr,i)   = [wd0(rr,i)**(1-gan_w4)]/[we0(rr,i)**(1-gan_w4)
+wd0(rr,i)**(1-gan_w4)];
alp_wz(rr,i)   = wz0(rr,i)/{(bet_we(rr,i)*[we0(rr,i)**gan_w4]
+bet_wd(rr,i)*[wd0(rr,i)**gan_w4])**(1/gan_w4)};

bet_wmm(rr,ss,i)  =  [(1+wtaum(rr,ss,i))*wmm0(rr,ss,i)**
(1-gan_wm)]/[sum(ww,(1+wtaum(rr,ww,i))*wmm0(rr,ww,i)**(1-g
an_wm))+(1+wtaumj(rr,i))*mj0(rr,i)**(1-gan_wm)];
bet_mj(rr,i)   = [(1+wtaumj(rr,i))*mj0(rr,i)**(1-gan_wm)]
/[sum(ww,(1+wtaum(rr,ww,i))*wmm0(rr,ww,i)**(1-gan_wm))+(1+
wtaumj(rr,i))*mj0(rr,i)**(1-gan_wm)];
alp_wmm(rr,i)  = wm0(rr,i)/{[sum(ss,bet_wmm(rr,ss,i)*wmm0(rr,ss,i
)**gan_wm)+bet_mj(rr,i)*mj0(rr,i)**gan_wm]**(1/gan_wm)};

bet_wm(rr,i)   = [wm0(rr,i)**(1-gan_w5)]/[wm0(rr,i)**(1-gan_w5)
+wd0(rr,i)**(1-gan_w5)];
bet_wd2(rr,i)  = [wd0(rr,i)**(1-gan_w5)]/[wm0(rr,i)**(1-gan_w5)
+wd0(rr,i)**(1-gan_w5)];
alp_wq(rr,i)   = wq0(rr,i)/{(bet_wm(rr,i)*[wm0(rr,i)**gan_w5]
+bet_wd2(rr,i)*[wd0(rr,i)**gan_w5])**(1/gan_w5)};

awx(rr,i,j) = wx0(rr,i,j)/wz0(rr,j);
awy(rr,j) = wy0(rr,j)/wz0(rr,j);

display
eneww,enecc,enewhh,enechh,eneccc,enechhh,beneww,benewhh,be
neccc,benechhh
```

VARIABLES
```
y(s,j)
l(s,j)
k(s,j)
py(s,j)
pl(s)
```

pk(s)
xx(r,i,s,j)
x(i,s,j)
px(i,s,j)
pq(r,i)
z(s,j)
pz(s,j)
xh(r,i,s)
xg(r,i,s)
xi(r,i,s)
tz(s,j)
tr(s)
epsilon
e(r,i)
d(r,i)
pe(i)
pd(r,i)
q(r,i)
pm(i)
m(r,i)
sp(s)
sg(s)
td(s)
pwmj(i)
pwej(i)
tez(r)
teh(r)
;

VARIABLES
wy(rr,j)
wlab(rr,j)
wlan(rr,j)
wk(rr,j)

```
wene(rr,j)
pwy(rr,j)
pwlab(rr)
pwlan(rr)
pwk(rr)
pwene(rr)
wx(rr,i,j)
pwq(rr,i)
wz(rr,j)
pwz(rr,j)
wxh(rr,i)
wxg(rr,i)
wxi(rr,i)
wtz(rr,j)
wepsilon(rr)
we(rr,i)
wd(rr,i)
pwe(rr,i)
pwd(rr,i)
ej(rr,i)
wee(rr,ss,i)
pwee(rr,ss,i)
pej(rr,i)
wmm(rr,ss,i)
mj(rr,i)
pwmm(rr,ss,i)
pmj(rr,i)
wq(rr,i)
wm(rr,i)
pwm(rr,i)
wsp(rr)
wsg(rr)
wtd(rr)
wpe(rr,ss,i)
```

```
wpm(rr,ss,i)
wpmj(rr,i)
wpej(rr,i)
wth(rr,i)
wtm(rr,ss,i)
wte(rr,ss,i)
wtmj(rr,i)
wtej(rr,i)
wtlab(rr,j)
wtk(rr,j)
wtlan(rr,j)
wtene(rr,j)
wtx(rr,i,j)
wtez(rr)
wteh(rr)
object;
```

EQUATION
```
eq_wy(rr,j),eq_wlab(rr,j),eq_wk(rr,j),eq_wlan(rr,j),eq_wen
e(rr,j),eq_tez(s),eq_teh(s)
eq_wz(rr,j),eq_wy2(rr,j),eq_wx(rr,i,j),eq_wxh(rr,i),eq_wxg
(rr,i),eq_wxi(rr,i)
eq_wtz(rr,j),eq_wz2(rr,i),eq_we(rr,i),eq_wd(rr,i),eq_we2(r
r,i),eq_wee(rr,ss,i)
eq_ej(rr,i),eq_wpee(rr,ss,i),eq_wpmm(rr,ss,i),eq_wm(rr,i),
eq_wmm(rr,ss,i)
eq_mj(rr,i),eq_wq(rr,i),eq_wm2(rr,i),eq_wd2(rr,i),eq_wtd(r
r),eq_wpej(rr,i)
eq_wpmj(rr,i),eq_wmkt(rr,i),eq_wflab(rr),eq_wflan(rr),eq_w
fk(rr),eq_wfene(rr),
eq_wsp(rr),eq_wsg(rr),eq_wsf(rr),eq_wth(rr,i),eq_wtm(rr,ss
,i),eq_wtmj(rr,i)
```

Appendix 179

```
eq_wte(rr,ss,i),eq_wtej(rr,i),eq_wtlab(rr,j),eq_wtk(rr,j),
eq_wtlan(rr,j),eq_wtene(rr,j)
eq_wtx(rr,i,j),eq_pwej(rr,i),eq_pwmj(rr,i),eq_y(s,j),eq_l(
s,j),eq_k(s,j)
eq_x(i,s,j),eq_xx(r,i,s,j),eq_z(s,j),eq_xz(i,s,j),eq_xh(r,
i,s),eq_xg(r,i,s)
eq_xi(r,i,s),eq_tz(s,j),eq_z1(r,i),eq_e1(r,i),eq_d1(r,i),e
q_q1(r,i),eq_m1(r,i)
eq_d12(r,i),eq_td(s),eq_pe(i),eq_pm(i),eq_mkt(r,i),eq_fl(s
),eq_fk(s),eq_sp(s)
eq_sg(s),eq_y2(s,j),eq_m_we(i),eq_e_wm(i),eq_wee_wmm(rr,ss
,i),
eq_wp(rr,ss,i),obj,eq_wee2(rr,ss,i),eq_z2(r,i),eq_z3(r,i),
eq_z4(r,i),eq_e2(r,i),
eq_e3(r,i),eq_d2(r,i),eq_d3(r,i),eq_q2(r,i),eq_q3(r,i),eq_
q4(r,i),eq_m2(r,i),
eq_m3(r,i),eq_d22(r,i),eq_d23(r,i),eq_y12(s,j),eq_y13(s,j)
,eq_y14(s,j),eq_l2(s,j),
eq_l3(s,j),eq_k2(s,j),eq_k3(s,j),eq_x2(i,s,j),eq_xx2(r,i,s
,j),eq_y21(s,j),eq_xz2(i,s,j),
eq_wy12(rr,j),eq_wy13(rr,j),eq_wlan12(rr,j),eq_wene2(rr,j)
,eq_wy21(rr,j),eq_wx2(rr,i,j),
eq_wmm2(rr,ss,i),eq_wxh2(rr,i),eq_wxg2(rr,i),eq_wxi2(rr,i)
,eq_xh2(r,i,s),eq_xg2(r,i,s),
eq_xi2(r,i,s),eq_wtez(rr),eq_wteh(rr),eq_sf,eq_tr2;

*


eq_y(s,j)$(l0(s,j) and  k0(s,j))..  y(s,j)  =e= alp_y(s,j)
*[bet_l(s,j)*l(s,j)**gan_1+bet_k(s,j)*k(s,j)**gan_1]**(1/g
an_1);
eq_y12(s,j)$(l0(s,j) and not k0(s,j))..  y(s,j) =e=  alp_y
(s,j)*l(s,j);
```

```
eq_y13(s,j)$(k0(s,j) and not l0(s,j)).. y(s,j) =e= alp_y
(s,j)*k(s,j);
eq_y14(s,j)$(not y0(s,j)).. y(s,j)   =e= 0;

eq_l(s,j)$(l0(s,j)and k0(s,j))..  l(s,j)*alp_y(s,j)   =e=  y(s,j)
*{py(s,j)*alp_y(s,j)*bet_l(s,j)/pl(s)}**[1/(1-gan_1)];
eq_l2(s,j)$(l0(s,j) and not k0(s,j)).. l(s,j)*pl(s)  =e=  y(s,j)
*py(s,j);
eq_l3(s,j)$(not l0(s,j)).. l(s,j)   =e= 0;

eq_k(s,j)$(l0(s,j)and k0(s,j))..  k(s,j)*alp_y(s,j)   =e=  y(s,j)
*{py(s,j)*alp_y(s,j)*bet_k(s,j)/pk(s)}**[1/(1-gan_1)];
eq_k2(s,j)$(k0(s,j) and not l0(s,j)).. k(s,j)*pk(s)  =e=  y(s,j)
*py(s,j);
eq_k3(s,j)$(not k0(s,j)).. k(s,j)   =e= 0;

eq_x(i,s,j)$x0(i,s,j).. x(i,s,j) =e= alp_x(i,s,j)*[sum (r,
bet_x(r,i,s,j)*xx(r,i,s,j)**gan_2)]**(1/gan_2);
eq_x2(i,s,j)$(not x0(i,s,j)).. x(i,s,j)  =e= 0;

eq_xx(r,i,s,j)$xx0(r,i,s,j).. xx(r,i,s,j)*alp_x(i,s,j) =e= x
(i,s,j)*{px(i,s,j)*alp_x(i,s,j)*bet_x(r,i,s,j)/[pq(r,i)+tc
o2(r,i)]}**(1/[1-gan_2]);
eq_xx2(r,i,s,j)$(not xx0(r,i,s,j)).. xx(r,i,s,j)   =e= 0;

eq_y2(s,j)$y0(s,j).. y(s,j)   =e=  ay(s,j)*z(s,j);
eq_y21(s,j)$(not y0(s,j)).. y(s,j)   =e=  0;

eq_xz(i,s,j)$(x0(i,s,j) gt 0).. x(i,s,j)  =e= ax(i,s,j)
*z(s,j);
eq_xz2(i,s,j)$(x0(i,s,j) = 0).. x(i,s,j)  =e=  0;

eq_z(s,j).. pz(s,j)  =e=  py(s,j)*ay(s,j)+sum(i,px(i,s,j)
*ax(i,s,j));
```

```
eq_xh(r,i,s)$bet_u(r,i,s)..  [pq(r,i)+tco2(r,i)]*xh (r,i,s)
=e=  bet_u(r,i,s)*[pl(s)*fl (s)+pk(s)*fk(s)-td(s) -sp(s)];
eq_xh2(r,i,s)$(not bet_u(r,i,s))..  xh(r,i,s)   =e= 0;

eq_xg(r,i,s)$bet_g(r,i,s)..  pq(r,i)*xg(r,i,s)  =e=  bet_g
(r,i,s)*[td(s)+sum(j,tz(s,j))+tez(r)+teh(r)-sg(s)];
eq_xg2(r,i,s)$(not bet_g(r,i,s))..  xg(r,i,s)   =e= 0;

eq_xi(r,i,s)$bet_i(r,i,s)..  pq(r,i)*xi(r,i,s)   =e=  bet_i
(r,i,s)*[sp(s)+sg(s)+tr(s)+sf(s)];
eq_xi2(r,i,s)$(not bet_i(r,i,s))..  xi(r,i,s)   =e=   0;

eq_tz(s,j)..  tz(s,j)   =e=  tauz(s,j)*pz(s,j)*z(s,j);

eq_z1(r,i)$(e0(r,i) and d0(r,i))..  z(r,i)   =e=  alp_z(r,i)
*[bet_e(r,i)*[e(r,i)**gan_4]+bet_d1(r,i)*[d(r,i)**gan_4]]*
*{1/gan_4};
eq_z2(r,i)$(d0(r,i) and not e0(r,i))..  z(r,i)   =e=  alp_z(r,i)
*d(r,i);
eq_z3(r,i)$(e0(r,i) and not d0(r,i))..  z(r,i)   =e=  alp_z(r,i)
*e(r,i);
eq_z4(r,i)$(not z0(r,i))..  z(r,i)   =e=   0;

eq_e1(r,i)$(e0(r,i) and d0(r,i))..  e(r,i)*alp_z(r,i)   =e=  z(r,i)
*{{(1+tauz(r,i))*pz(r,i)*alp_z(r,i)*bet_e(r,i)/pe(i)}**[1/
(1-gan_4)]};
eq_e2(r,i)$(e0(r,i)  and  not  d0(r,i))..   pe(i)*e(r,i)   =e=
(1+tauz(r,i))*pz(r,i)*z(r,i);
eq_e3(r,i)$(e0(r,i) = 0)..  e(r,i)   =e=  0;

eq_d1(r,i)$(d0(r,i)  and  e0(r,i))..   d(r,i)*alp_z(r,i)   =e=
z(r,i)*{{(1+tauz(r,i))*pz(r,i)*alp_z(r,i)*bet_d1(r,i)/pd(r
,i)}**[1/(1-gan_4)]};
```

```
eq_d2(r,i)$(d0(r,i) and not e0(r,i))..  pd(r,i)*d(r,i)  =e=
(1+tauz(r,i))*pz(r,i)*z(r,i);
eq_d3(r,i)$(d0(r,i) = 0)..  d(r,i)  =e=  0;

eq_q1(r,i)$(m0(r,i) and d0(r,i))..  q(r,i)  =e=  alp_q(r,i)
*[bet_m(r,i)*m(r,i)**gan_5+bet_d2(r,i)*d(r,i)**gan_5]**{1/
gan_5};
eq_q2(r,i)$(d0(r,i) and not m0(r,i))..  q(r,i)  =e=  alp_q(r,i)
*d(r,i);
eq_q3(r,i)$(m0(r,i) and not d0(r,i))..  q(r,i)  =e=  alp_q(r,i)
*m(r,i);
eq_q4(r,i)$(not q0(r,i))..  q(r,i)  =e=  0;

eq_m1(r,i)$(m0(r,i) and d0(r,i))..  m(r,i)*alp_q(r,i)  =e=  q(r,i)
*{{pq(r,i)*alp_q(r,i)*bet_m(r,i)/pm(i)}**[1/(1-gan_5)]};
eq_m2(r,i)$(not m0(r,i))..  m(r,i)  =e=  0;
eq_m3(r,i)$(m0(r,i) and not d0(r,i))..  pm(i)*m(r,i)  =e=
pq(r,i)*q(r,i);

eq_d12(r,i)$(d0(r,i) and m0(r,i))..  d(r,i)*alp_q(r,i)  =e=
q(r,i)*{{pq(r,i)*alp_q(r,i)*bet_d2(r,i)/pd(r,i)}**{1/(1-ga
n_5)}};
eq_d22(r,i)$(d0(r,i) and not m0(r,i))..  pd(r,i)*d(r,i)  =e=
pq(r,i)*q(r,i);
eq_d23(r,i)$(not d0(r,i))..  d(r,i)  =e=  0;

eq_sf(r)..  sf(r)  =e=  sum(i,pwmj(i)*m(r,i))-sum(i,pwej(i)
*e(r,i));

eq_td(s)..  td(s)  =e=  taud*[pl(s)*fl(s)+pk(s)*fk(s)];

eq_pe(i)..  pe(i)  =e=  epsilon*pwej(i);

eq_pm(i)..  pm(i)  =e=  epsilon*pwmj(i);
```

eq_mkt(r,i)..   q(r,i) =e= sum(s,xh(r,i,s))+sum((s,j),xx(r,i,s,j))+sum(s,xg(r,i,s))+sum(s,xi(r,i,s));

eq_fl(s).. sum(j,l(s,j)) =e= fl(s);

eq_fk(s).. sum(j,k(s,j)) =e= fk(s);

eq_sp(s).. sp(s) =e= ps(s)*[pl(s)*fl(s)+pk(s)*fk(s)];

eq_sg(s)..   sg(s) =e= pg(s)*[td(s)+sum(j,tz(s,j))+tez(s)+teh(s)];

eq_tez(r).. tez(r) =e= sum((i,s,j),tco2(r,i)*xx (r,i,s,j));

eq_teh(r).. teh(r) =e= sum((i,s),tco2(r,i)*xh(r,i,s));

eq_tr2.. sum(s,tr(s)) =e= 0;

eq_wy(rr,j)$(wlab0(rr,j) and wk0(rr,j) and wlan0(rr,j) and wene0(rr,j)).. wy(rr,j) =e= alp_wy(rr,j)*[bet_wlab(rr,j)*wlab(rr,j)**gan_w1+bet_wk(rr,j)*wk(rr,j)**gan_w1+bet_wlan(rr,j)*wlan(rr,j)**gan_w1+bet_wene(rr,j)*wene(rr,j)**gan_w1]**(1/gan_w1);
eq_wy12(rr,j)$(wlab0(rr,j) and wk0(rr,j) and not wlan0(rr,j) and wene0(rr,j)).. wy(rr,j) =e= alp_wy(rr,j)*[bet_wlab(rr,j)*wlab(rr,j)**gan_w1+bet_wk(rr,j)*wk(rr,j)**gan_w1+bet_wene(rr,j)*wene(rr,j)**gan_w1]**(1/gan_w1);
eq_wy13(rr,j)$(wlab0(rr,j) and wk0(rr,j) and not wlan0(rr,j) and not wene0(rr,j)).. wy(rr,j) =e=   alp_wy(rr,j)*[bet_

wlab(rr,j)*wlab(rr,j)**gan_w1+bet_wk(rr,j)*wk(rr,j)**gan_w1]**(1/gan_w1);

eq_wlab(rr,j).. wlab(rr,j)*alp_wy(rr,j) =e= wy(rr,j) *{pwy(rr,j)*alp_wy(rr,j)*bet_wlab(rr,j)/[(1+wtaulab(rr,j))*pwlab(rr)]}**[1/(1-gan_w1)];

eq_wlan(rr,j)$wlan0(rr,j).. wlan(rr,j)*alp_wy(rr,j) =e= wy(rr,j)*{pwy(rr,j)*alp_wy(rr,j)*bet_wlan(rr,j)/[(1+wtaulan(rr,j))*pwlan(rr)]}**[1/(1-gan_w1)];
eq_wlan12(rr,j)$(not wlan0(rr,j)).. wlan(rr,j) =e= 0;

eq_wk(rr,j).. wk(rr,j)*alp_wy(rr,j) =e= wy(rr,j)*{pwy(rr,j)*alp_wy(rr,j)*bet_wk(rr,j)/[(1+wtauk(rr,j))*pwk(rr)]}**[1/(1-gan_w1)];

eq_wene(rr,j)$wene0(rr,j).. wene(rr,j)*alp_wy(rr,j) =e= wy(rr,j)*{pwy(rr,j)*alp_wy(rr,j)*bet_wene(rr,j)/[(1+wtauene(rr,j))*pwene(rr)]}**[1/(1-gan_w1)];
eq_wene2(rr,j)$(not wene0(rr,j)).. wene(rr,j) =e= 0;

eq_wy2(rr,j)$wy0(rr,j).. wy(rr,j) =e= awy(rr,j)*wz(rr,j);
eq_wy21(rr,j)$(not wy0(rr,j)).. wy(rr,j) =e= 0;

eq_wx(rr,i,j)$wx0(rr,i,j).. wx(rr,i,j) =e= awx(rr,i,j)*wz(rr,j);
eq_wx2(rr,i,j)$(not wx0(rr,i,j)).. wx(rr,i,j) =e= 0;

eq_wz(rr,j).. pwz(rr,j) =e= pwy(rr,j)*awy(rr,j)+sum(i, (pwq(rr,i)+wtaux(rr,i,j)*pwq(rr,i)+twco2(rr,i))*awx(rr,i,j));

eq_wxh(rr,i)$wxh0(rr,i).. (pwq(rr,i)+wtauh(rr,i)*pwq(rr,i)+twco2(rr,i))*wxh(rr,i) =e= bet_wxh(rr,i)*[pwlab(rr)*wflab

```
(rr)+pwlan(rr)*wflan(rr)+pwk(rr)*wfk(rr)+pwene(rr)*wfene(r
r)-wtd(rr)-wsp(rr)];
eq_wxh2(rr,i)$(not wxh0(rr,i)).. wxh(rr,i)  =e=  0;

eq_wxg(rr,i)$wxg0(rr,i)..  pwq(rr,i)*wxg(rr,i)  =e=  bet_wxg
(rr,i)*[wtd(rr)+sum(j,wtz(rr,j)+wth(rr,j)+wtlan(rr,j)+wtk(
rr,j)+wtene(rr,j)+wtlab(rr,j)+wtmj(rr,j)+wtej(rr,j))+sum((
ss,j),wte(rr,ss,j)+wtm(rr,ss,j))+sum((h,j),wtx(rr,h,j))+wt
ez(rr)+wteh(rr)-wsg(rr)];
eq_wxg2(rr,i)$(not wxg0(rr,i)).. wxg(rr,i)  =e=  0;

eq_wxi(rr,i)$wxi0(rr,i)..  pwq(rr,i)*wxi(rr,i)   =e=  bet_wxi
(rr,i)*[wsp(rr)+wsg(rr)+wepsilon(rr)*wsf(rr)];
eq_wxi2(rr,i)$(not wxi0(rr,i)).. wxi(rr,i)  =e=  0;

eq_wtz(rr,j).. wtz(rr,j) =e= wtauz(rr,j)*pwz(rr,j) *wz(rr,j);

eq_wz2(rr,i)..   wz(rr,i) =e=  alp_wz(rr,i)*[bet_we(rr,i)*we
(rr,i)**gan_w4+bet_wd(rr,i)*wd(rr,i)**gan_w4]**{1/gan_w4};

eq_we(rr,i)..  we(rr,i)*alp_wz(rr,i)    =e=   wz(rr,i)*{{(1+
wtauz(rr,i))*pwz(rr,i)*alp_wz(rr,i)*bet_we(rr,i)/pwe(rr,i)
}**[1/(1-gan_w4)]};

eq_wd(rr,i)..   wd(rr,i)*alp_wz(rr,i)   =e=   wz(rr,i)*{{(1+
wtauz(rr,i))*pwz(rr,i)*alp_wz(rr,i)*bet_wd(rr,i)/pwd(rr,i)
}**[1/(1-gan_w4)]};

eq_we2(rr,i)..   we(rr,i)   =e=   alp_wee(rr,i)*[sum(ss,bet_wee
(rr,ss,i)*wee(rr,ss,i)**gan_we)+bet_ej(rr,i)*ej(rr,i)**gan
_we]**{1/gan_we};

eq_wee(rr,ss,i)$wee0(rr,ss,i)..    wee(rr,ss,i)*alp_wee(rr,i)
=e=  we(rr,i)*{{pwe(rr,i)*alp_wee(rr,i)*bet_wee(rr,ss,i)
```

```
/[(1-wtaue(rr,ss,i))*pwee(rr,ss,i)]}**[1/(1-gan_we)]};
eq_wee2(rr,ss,i)$(not wee0(rr,ss,i)).. wee(rr,ss,i) =e= 0;

eq_ej(rr,i)..   ej(rr,i)*alp_wee(rr,i)  =e=  we(rr,i)*{{pwe
(rr,i)*alp_wee(rr,i)*bet_ej(rr,i)/[(1-wtauej(rr,i))*pej(rr
,i)]}**[1/(1-gan_we)]};

eq_wm2(rr,i)..  wm(rr,i) =e= alp_wmm(rr,i)*[sum(ss, bet_wmm
(rr,ss,i)*wmm(rr,ss,i)**gan_wm)+bet_mj(rr,i)*mj(rr,i)**gan
_wm]**{1/gan_wm};

eq_wmm(rr,ss,i)$wmm0(rr,ss,i).. wmm(rr,ss,i)*alp_wmm(rr,i)
=e= wm(rr,i)*{{pwm(rr,i)*alp_wmm(rr,i)*bet_wmm(rr,ss,i)
/[(1+wtaum(rr,ss,i))*pwmm(rr,ss,i)]}**[1/(1-gan_wm)]};
eq_wmm2(rr,ss,i)$(not wmm0(rr,ss,i)).. wmm(rr,ss,i) =e= 0;

eq_mj(rr,i)..   mj(rr,i)*alp_wmm(rr,i)  =e=  wm(rr,i)*{{pwm
(rr,i)*alp_wmm(rr,i)*bet_mj(rr,i)/[(1+wtaumj(rr,i))*pmj(rr
,i)]}**[1/(1-gan_wm)]};

eq_wq(rr,i)..   wq(rr,i)  =e=  alp_wq(rr,i)*[bet_wm(rr,i)*wm
(rr,i)**gan_w5+bet_wd2(rr,i)*wd(rr,i)**gan_w5]**{1/gan_w5}
;

eq_wm(rr,i).. wm(rr,i)*alp_wq(rr,i)  =e=  wq(rr,i)*{{pwq (rr,i)
*alp_wq(rr,i)*bet_wm(rr,i)/pwm(rr,i)}**[1/(1-gan_w5)]};

eq_wd2(rr,i)..  wd(rr,i)*alp_wq(rr,i)   =e=   wq(rr,i)*{{pwq
(rr,i)*alp_wq(rr,i)*bet_wd2(rr,i)/pwd(rr,i)}**{1/(1-gan_w5
)}};

eq_wsf(rr).. wsf(rr)  =e=  sum((ss,i),wpm(rr,ss,i)*wmm(rr,ss,
i))+sum(i,wpmj(rr,i)*mj(rr,i))-sum((ss,i),wpe(rr,ss,i)*wee
(rr,ss,i))-sum(i,wpej(rr,i)*ej(rr,i));
```

eq_wtd(rr).. wtd(rr) =e= wtaud(rr)*[pwlab(rr)*wflab(rr) +pwlan
rr)*wflan(rr)+pwk(rr)*wfk(rr)+pwene(rr)*wfene(rr)];

eq_wpee(rr,ss,i).. pwee(rr,ss,i) =e= wepsilon(rr)*wpe
(rr,ss,i);

eq_wpmm(rr,ss,i).. pwmm(rr,ss,i) =e= wepsilon(rr)*wpm
(rr,ss,i);
eq_wpej(rr,i).. pej(rr,i) =e= wepsilon(rr)*wpej(rr,i);
eq_wpmj(rr,i).. pmj(rr,i) =e= wepsilon(rr)*wpmj(rr,i);
eq_wp(rr,ss,i).. wpe(rr,ss,i)-wpm(ss,rr,i) =e= 0;
eq_pwej(rr,i).. wpej(rr,i) =e= pwmj(i);
eq_pwmj(rr,i).. wpmj(rr,i) =e= pwej(i);
eq_wmkt(rr,i).. wq(rr,i) =e= wxh(rr,i)+sum(j,wx(rr,i,j))
+wxg(rr,i)+wxi(rr,i);
eq_wflab(rr).. sum(j,wlab(rr,j)) =e= wflab(rr);
eq_wfk(rr).. sum(j,wk(rr,j)) =e= wfk(rr);
eq_wflan(rr).. sum(j,wlan(rr,j)) =e= wflan(rr);
eq_wfene(rr).. sum(j,wene(rr,j)) =e= wfene(rr);
eq_wsp(rr).. wsp(rr) =e= wps(rr)*(pwlab(rr)*wflab(rr) +pwlan
(rr)*wflan(rr)+pwk(rr)*wfk(rr)+pwene(rr)*wfene(rr));
eq_wsg(rr).. wsg(rr) =e= wpg(rr)*[wtez(rr)+wteh(rr)+wtd
(rr)+sum(j,wtz(rr,j)+wth(rr,j)+wtlan(rr,j)+wtk(rr,j)+wtene
(rr,j)+wtlab(rr,j)+wtmj(rr,j)+wtej(rr,j))+sum((ss,j),wte(r
r,ss,j)+wtm(rr,ss,j))+sum((h,j),wtx(rr,h,j))];
eq_wth(rr,i).. wth(rr,i) =e= wtauh(rr,i)*pwq(rr,i)*wxh (rr,i);
eq_wtm(rr,ss,i).. wtm(rr,ss,i) =e= wtaum(rr,ss,i)*pwmm(rr,
ss,i)*wmm(rr,ss,i);
eq_wtmj(rr,i).. wtmj(rr,i) =e= wtaumj(rr,i)*pmj(rr,i)*mj
(rr,i);
eq_wte(rr,ss,i).. wte(rr,ss,i) =e= wtaue(rr,ss,i)*pwee(rr,
ss,i)*wee(rr,ss,i);

```
eq_wtej(rr,i).. wtej(rr,i) =e= wtauej(rr,i)*pej(rr,i)*ej
(rr,i);
eq_wtlab(rr,j).. wtlab(rr,j) =e= wtaulab(rr,j)*pwlab(rr)
*wlab(rr,j);
eq_wtk(rr,j).. wtk(rr,j) =e= wtauk(rr,j)*pwk(rr)*wk(rr,j);
eq_wtlan(rr,j).. wtlan(rr,j) =e= wtaulan(rr,j)*pwlan(rr)
*wlan(rr,j);
eq_wtene(rr,j).. wtene(rr,j) =e= wtauene(rr,j)*pwene(rr)
*wene(rr,j);
eq_wtx(rr,i,j).. wtx(rr,i,j) =e= wtaux(rr,i,j)*pwq(rr,i)
*wx(rr,i,j);
eq_m_we(i).. sum(r,e(r,i)) =e= sum(rr,mj(rr,i));
eq_e_wm(i).. sum(r,m(r,i)) =e= sum(rr,ej(rr,i));
eq_wee_wmm(rr,ss,i).. wee(rr,ss,i)-wmm(ss,rr,i) =e= 0;
eq_wtez(rr).. wtez(rr) =e= sum((i,j),twco2(rr,i)*wx(rr,
i,j));
eq_wteh(rr).. wteh(rr) =e= sum(i,twco2(rr,i)*wxh(rr,i));
obj.. object =e= epsilon;
```

*INITIALIZING VARIABLES--

```
y.l(s,j)     = y0(s,j);
l.l(s,j)     = l0(s,j);
k.l(s,j)     = k0(s,j);
xx.l(r,i,s,j) = xx0(r,i,s,j);
x.l(i,s,j)   = x0(i,s,j);
z.l(s,j)     = z0(s,j);
xh.l(r,i,s)  = xh0(r,i,s);
tz.l(s,j)    = tz0(s,j);
epsilon.l    = 1;
e.l(r,i)     = e0(r,i);
d.l(r,i)     = d0(r,i);
q.l(r,i)     = q0(r,i);
m.l(r,i)     = m0(r,i);
tr.l(s)      = tr0(s);
td.l(s)      = td0(s);
```

```
sp.l(s)       = sp0(s);
sg.l(s)       = sg0(s);
xg.l(r,i,s)   = xg0(r,i,s);
xi.l(r,i,s)   = xi0(r,i,s);
wy.l(rr,j)    = wy0(rr,j);
wlab.l(rr,j)  = wlab0(rr,j);
wlan.l(rr,j)  = wlan0(rr,j);
wene.l(rr,j)  = wene0(rr,j);
wk.l(rr,j)    = wk0(rr,j);
wx.l(rr,i,j)  = wx0(rr,i,j);
wz.l(rr,j)    = wz0(rr,j);
wxh.l(rr,i)   = wxh0(rr,i);
wtz.l(rr,j)   = wtz0(rr,j);
wepsilon.l(rr) = 1;
wee.l(rr,ss,i) = wee0(rr,ss,i);
ej.l(rr,i)    = ej0(rr,i);
we.l(rr,i)    = we0(rr,i);
wd.l(rr,i)    = wd0(rr,i);
wq.l(rr,i)    = wq0(rr,i);
wmm.l(rr,ss,i) = wmm0(rr,ss,i);
mj.l(rr,i)    = mj0(rr,i);
wm.l(rr,i)    = wm0(rr,i);
wtd.l(rr)     = wtd0(rr);
wsp.l(rr)     = wsp0(rr);
wsg.l(rr)     = wsg0(rr);
wxg.l(rr,i)   = wxg0(rr,i);
wxi.l(rr,i)   = wxi0(rr,i);
wth.l(rr,i)   = wth0(rr,i);
wtm.l(rr,ss,i) = wtm0(rr,ss,i);
wte.l(rr,ss,i) = wte0(rr,ss,i);
wtmj.l(rr,i)  = wtmj0(rr,i);
wtej.l(rr,i)  = wtej0(rr,i);
wtlab.l(rr,j) = wtlab0(rr,j);
wtk.l(rr,j)   = wtk0(rr,j);
```

```
wtlan.l(rr,j)   = wtlan0(rr,j);
wtene.l(rr,j)   = wtene0(rr,j);
wtx.l(rr,i,j)   = wtx0(rr,i,j);
py.l(s,j)    = 1;
pl.l(s)      = 1;
pk.l(s)      = 1;
px.l(i,s,j)  = 1;
pq.l(r,i)    = 1;
pz.l(s,j)    = 1;
pe.l(i)      = 1;
pd.l(r,i)    = 1;
pm.l(i)      = 1;
pwy.l(rr,j)  = 1;
pwlab.l(rr)  = 1;
pwlan.l(rr)  = 1;
pwk.l(rr)    = 1;
pwene.l(rr)  = 1;
pwq.l(rr,i)  = 1;
pwz.l(rr,i)  = 1;
pwe.l(rr,i)  = 1;
pwee.l(rr,ss,i)  = 1;
pwd.l(rr,i)  = 1;
pwm.l(rr,i)  = 1;
pwmm.l(rr,ss,i)  = 1;
pwej.l(i)    = 1;
pwmj.l(i)    = 1;
wpe.l(rr,ss,i)   = 1;
wpm.l(rr,ss,i)   = 1;
pej.l(rr,i)  = 1;
pmj.l(rr,i)  = 1;
wpmj.l(rr,i) = 1;
wpej.l(rr,i) = 1;
```

*SETTING LOWER BOUNDS TO AVOID DIVISION BY ZERO--------------------
```
$ontext
y.lo(s,j)      = 0;
l.lo(s,j)      = 0;
k.lo(s,j)      = 0;
xx.lo(r,i,s,j) = 0;
x.lo(i,s,j)    = 0;
z.lo(s,j)      = 0;
xh.lo(r,i,s)   = 0;
epsilon.lo     = 0;
e.lo(r,i)      = 0;
d.lo(r,i)      = 0;
q.lo(r,i)      = 0;
m.lo(r,i)      = 0;
xg.lo(r,i,s)   = 0;
xi.lo(r,i,s)   = 0;
sp.lo(s)       = 0;
tez.lo(r)      = 0;
teh.lo(r)      = 0;
wtez.lo(rr)    = 0;
wteh.lo(rr)    = 0;
wy.lo(rr,j)    = 0;
wlab.lo(rr,j)  = 0;
wlan.lo(rr,j)  = 0;
wk.lo(rr,j)    = 0;
wene.lo(rr,j)  = 0;
wx.lo(rr,i,j)  = 0;
wz.lo(rr,j)    = 0;
wxh.lo(rr,i)   = 0;
wepsilon.lo(rr) = 0;
we.lo(rr,i)    = 0;
wd.lo(rr,i)    = 0;
wq.lo(rr,i)    = 0;
wm.lo(rr,i)    = 0;
```

```
wmm.lo(rr,ss,i)   = 0;
wee.lo(rr,ss,i)   = 0;
wxg.lo(rr,i)   = 0;
wxi.lo(rr,i)   = 0;
wsp.lo(rr)   = 0;
pwy.lo(rr,j)   = 0;
pwlab.lo(rr)   = 0;
pwlan.lo(rr)   = 0;
pwk.lo(rr)   = 0;
pwene.lo(rr)   = 0;
pwq.lo(rr,i)   = 0;
pwz.lo(rr,j)   = 0;
pwe.lo(rr,i)   = 0;
pwd.lo(rr,i)   = 0;
pwm.lo(rr,i)   = 0;
pwej.lo(i)   = 0;
pwmj.lo(i)   = 0;
wpej.lo(rr,i)   = 0;
wpmj.lo(rr,i)   = 0;
wpe.lo(rr,ss,i)   = 0;
wpm.lo(rr,ss,i)   = 0;
pwee.lo(rr,ss,i)   = 0;
pwmm.lo(rr,ss,i)   = 0;
pej.lo(rr,i)   = 0;
pmj.lo(rr,i)   = 0;
py.lo(s,j)   = 0;
pl.lo(s)   = 0;
pk.lo(s)   = 0;
px.lo(i,s,j)   = 0;
pq.lo(r,i)   = 0;
pz.lo(s,j)   = 0;
pe.lo(i)   = 0;
pd.lo(r,i)   = 0;
pm.lo(i)   = 0;
```

```
$offtext


$ontext
y.fx(s,j)      = y0(s,j);
l.fx(s,j)      = l0(s,j);
k.fx(s,j)      = k0(s,j);
xx.fx(r,i,s,j) = xx0(r,i,s,j);
x.fx(i,s,j)    = x0(i,s,j);
z.fx(s,j)      = z0(s,j);
xh.fx(r,i,s)   = xh0(r,i,s);
tz.fx(s,j)     = tz0(s,j);
e.fx(r,i)      = e0(r,i);
d.fx(r,i)      = d0(r,i);
q.fx(r,i)      = q0(r,i);
m.fx(r,i)      = m0(r,i);
tr.fx(s)       = tr0(s);
td.fx(s)       = td0(s);
sp.fx(s)       = sp0(s);
sg.fx(s)       = sg0(s);
xg.fx(r,i,s)   = xg0(r,i,s);
xi.fx(r,i,s)   = xi0(r,i,s);
wy.fx(rr,j)    = wy0(rr,j);
wlab.fx(rr,j)  = wlab0(rr,j);
wlan.fx(rr,j)  = wlan0(rr,j);
wene.fx(rr,j)  = wene0(rr,j);
wk.fx(rr,j)    = wk0(rr,j);
wx.fx(rr,i,j)  = wx0(rr,i,j);
wz.fx(rr,j)    = wz0(rr,j);
wxh.fx(rr,i)   = wxh0(rr,i);
wtz.fx(rr,j)   = wtz0(rr,j);
wepsilon.fx(rr) = 1;
wee.fx(rr,ss,i) = wee0(rr,ss,i);
ej.fx(rr,i)    = ej0(rr,i);
```

```
we.fx(rr,i)     = we0(rr,i);
wd.fx(rr,i)     = wd0(rr,i);
wq.fx(rr,i)     = wq0(rr,i);
wmm.fx(rr,ss,i) = wmm0(rr,ss,i);
mj.fx(rr,i)     = mj0(rr,i);
wm.fx(rr,i)     = wm0(rr,i);
wtd.fx(rr)      = wtd0(rr);
wsp.fx(rr)      = wsp0(rr);
wsg.fx(rr)      = wsg0(rr);
wxg.fx(rr,i)    = wxg0(rr,i);
wxi.fx(rr,i)    = wxi0(rr,i);
wth.fx(rr,i)    = wth0(rr,i);
wtm.fx(rr,ss,i) = wtm0(rr,ss,i);
wte.fx(rr,ss,i) = wte0(rr,ss,i);
wtmj.fx(rr,i)   = wtmj0(rr,i);
wtej.fx(rr,i)   = wtej0(rr,i);
wtlab.fx(rr,j)  = wtlab0(rr,j);
wtk.fx(rr,j)    = wtk0(rr,j);
wtlan.fx(rr,j)  = wtlan0(rr,j);
wtene.fx(rr,j)  = wtene0(rr,j);
wtx.fx(rr,i,j)  = wtx0(rr,i,j);
pwy.fx(rr,j)    = 1;
pwlab.fx(rr)    = 1;
pwlan.fx(rr)    = 1;
pwk.fx(rr)      = 1;
pwene.fx(rr)    = 1;
pwq.fx(rr,i)    = 1;
pwz.fx(rr,j)    = 1;
pwe.fx(rr,i)    = 1;
pwd.fx(rr,i)    = 1;
pwm.fx(rr,i)    = 1;
pwej.fx(i)      = 1;
pwmj.fx(i)      = 1;
wpej.fx(rr,i)   = 1;
```

```
wpmj.fx(rr,i)     = 1;
wpe.fx(rr,ss,i)   = 1;
wpm.fx(rr,ss,i)   = 1;
pwee.fx(rr,ss,i)  = 1;
pwmm.fx(rr,ss,i)  = 1;
pej.fx(rr,i)      = 1;
pmj.fx(rr,i)      = 1;
py.fx(s,j)        = 1;
pl.fx(s)          = 1;
pk.fx(s)          = 1;
px.fx(i,s,j)      = 1;
pq.fx(r,i)        = 1;
pz.fx(s,j)        = 1;
pe.fx(i)          = 1;
pd.fx(r,i)        = 1;
pm.fx(i)          = 1;
$offtext

pl.fx("noea")   =   1;
pwlab.fx(rr)    =   1;
*wepsilon.fx("ANZ") =   1;
```

*DEFINING AND SOLVE THE MODEL--

```
MODEL project23 /ALL/;
project23.iterlim = 10000;
option reslim = 10000000;
option NLP = CONOPT;
SOLVE project23 MAXIMIZING object USING NLP;
```

Model 3: One National, Multi-Region, Dynamic Model

```
$ Title  Recursive Dynamic Multiregional CGE Model for China
Economy

option solprint = off;
```

*==
*   Read data
*==

```
set u/noea,nomu,noco,eaco,soco,cent,nowe,sowe,


agr,cm,ogm,om,fm,tex,wa,swp,pp,ppc,ci,nmmp,msp,mp,mi,te,em
e,ece,omi,ewgs,cons,tw,com,ser,

l,k,pt,dt,hoh,gov,inv,ex,im,ops/

r(u)/noea,nomu,noco,eaco,soco,cent,nowe,sowe/

i(u)/
agr,cm,ogm,om,fm,tex,wa,swp,pp,ppc,ci,nmmp,msp,mp,mi,te,em
e,ece,omi,ewgs,cons,tw,com,ser/



alias (U,V),(R,S),(I,J);


*LOADING DATA
TABLE I_M(R,I,S,J)
$INCLUDE IntermediateInput.INC
```

;

```
TABLE F_D(R,I,S,V)
$INCLUDE FinalDemand.INC
;

TABLE V_A(U,r,J)
$INCLUDE ValueAdded.INC
;

TABLE EXT(R,I,V)
$INCLUDE Trade.INC
;
```

Parameters

y0(j,r)	Output
yd0(j,r)	Output for domestic use
ya0(j,r)	Armington aggregates
ijs0(i,j,s,r)	Intermediate inputs
labs0(j,r)	Labor input inputs
lab0(r)	Labor supply
caps0(j,r)	Capital inputs
cap0(r)	Capital supply
ty0(j,r)	Indirect tax on output
cps0(i,s,r)	Private consumption
cp0(r)	Private consumption
cgs0(i,s,r)	Government consumption
cg0(r)	Government consumption
invs0(i,s,r)	Private capital formation
inv0(r)	Private capital formation
ex0(i,r)	Exports
im0(i,r)	Imports
tm0(i,r)	Tariff

;

```
ijs0(i,j,s,r)  = I_M(R,I,S,J)/100;
labs0(j,r)     = V_A("L",r,J)/100;
caps0(j,r)     = V_A("K",r,J)/100;
ty0(j,r)       = V_A("PT",r,J)/100;
cps0(i,s,r)    = F_D(R,I,S,"HOH")/100;
cgs0(i,s,r)    = F_D(R,I,S,"GOV")/100;
invs0(i,s,r)   = F_D(R,I,S,"INV")/100;
ex0(i,r)       = EXT(R,I,"EX")/100;
im0(i,r)       = EXT(R,I,"IM")/100;
tm0(i,r)       = 0;


y0(j,r)   =   sum((i,s),ijs0(i,j,s,r))+labs0(j,r)+caps0(j,r)
+ty0(j,r);

yd0(j,r)  =   y0(j,r)-ex0(j,r);
ya0(j,r)  =   yd0(j,r)-im0(j,r)-tm0(j,r);


cp0(r)    =   sum((i,s),cps0(i,s,r));
cg0(r)    =   sum((i,s),cgs0(i,s,r));

inv0(r)   =   sum((i,s),invs0(i,s,r));


*================================================================
*      Tax
*================================================================


Parameter
    taxy0(j,r)        Output tax rate
    taxl0(r)          labor income tax rate
    taxk0(r)          Capital income tax rate
    taxm0(j,r)        Tariff rate
```

```
    taxe0(j,r)
    pa0(j,r)
    py0(j,r)           Output price tax excluded
    pl0(r)             Labor cost tax included
    pk0(r)             Capital cost tax included
    pm0(j,r)           Import price tax included
;

taxe0(j,r)  =  0;
pa0(j,r)    =  1+taxe0(j,r);
taxy0(j,r)  =  ty0(j,r)/y0(j,r);
py0(j,r)    =  1-taxy0(j,r);

taxl0(r)    =  0.15;
taxk0(r)    =  0.15;




labs0(j,r)  =  labs0(j,r)*(1-taxl0(r));
lab0(r)     =  sum(j,labs0(j,r));
taxl0(r)    =  taxl0(r)/(1-taxl0(r));
pl0(r)      =  1+taxl0(r);

caps0(j,r)  =  caps0(j,r)*(1-taxk0(r));
cap0(r)     =  sum(j,caps0(j,r));
taxk0(r)    =  taxk0(r)/(1-taxk0(r));
pk0(r)      =  1+taxk0(r);

taxm0(j,r)$im0(j,r)   =  tm0(j,r)/im0(j,r);
pm0(j,r)    =  1+taxm0(j,r);

*===============================================================
*    Parameter for dynamic model
```

```
scalar
    esubt0      Intertemporal elasticity of substitution    /0.5/
    depr0       Annual depreciation rate                    /0.04/
    r0          Benchmark interest rate                     /0.05/;

set
    t           Time periods                                /1997/,
    t0(t)       First period of the model
    tl(t)       Final period of the model

t0(t) = yes$(ord(t) eq 1);
tl(t) = yes$(ord(t) eq card(t));

table gr0(*,*)  Growth rate
```

	noea	nomu	noco	eaco	soco	cent	nowe	sowe
1997	13	13	13	13	13	13	13	13
1998	13	13	13	13	13	13	13	13
1999	13	13	13	13	13	13	13	13
2000	13	13	13	13	13	13	13	13
2001	13	13	13	13	13	13	13	13
2002	13	13	13	13	13	13	13	13
2003	13	13	13	13	13	13	13	13
2004	13	13	13	13	13	13	13	13
2005	13	13	13	13	13	13	13	13
2006	13	13	13	13	13	13	13	13
2007	13	13	13	13	13	13	13	13
2008	13	13	13	13	13	13	13	13
2009	13	13	13	13	13	13	13	13
2010	13	13	13	13	13	13	13	13
2011	13	13	13	13	13	13	13	13
2012	13	13	13	13	13	13	13	13

```
;

gr0(t,r) = gr0(t,r)/100;
```

Parameter
    year(t)   Year
    qref(t,r)   Reference quantity path
    pref(t,r)   Reference path of present value prices

```
year(t) = 1997+(ord(t)-1);
qref(t,r) = 1;
loop(t, qref(t+1,r) = qref(t,r)*(1+gr0(t,r)););
pref(t,r) = (1/(1+r0))**(ord(t)-1);
```

*==
*   Calibration of savings and investment to a steady-state
*   equilibrium across all regions:
*==

Parameters
    rk0(r)   Price of capital service
    rktx0(r)   Price of capital service tax include
    pk0(r)   Benchmark capital price
    k0(r)   Base year capital
    vk0(r)   Value of capital earnings
    vi0(r)   Value of investment on steady state
    vi00(r)   Value of investment observed
    lammda(r)   Scaling factor for Adjustment
    c0adj(r)   Increment in consumption for investment
    i0adj(r)   Increment in investment to match benchmark
    vtax(r)   Implicit tax

```
rk0(r) = r0+depr0;
```

```
rktx0(r)   =   rk0(r)*(1+taxk0(r));
pk0(r)     =   1+r0;


vk0(r)     =   sum(j,caps0(j,r));
k0(r)      =   vk0(r)/rk0(r);


vi0(r)     =   (gr0("1997",r)+depr0)*k0(r);

vi00(r)    =   inv0(r);

i0adj(r)   =   max(0,vi0(r)-vi00(r));
c0adj(r)   =   max(0,vi00(r)-vi0(r));


*display cps0,invs0;
display cp0,inv0;
```

```
*===============================================================
*      Saving
*===============================================================

Parameters
    hsave0(r)        Household saving
    gsave0(r)        Government saving
    bop0(r)          Balance of payment
    trn0(*)          Transfer beween region
;
hsave0(r)   =   lab0(r)+cap0(r)-cp0(r);
gsave0(r)   =   sum(j,ty0(j,r))+lab0(r)*tax10(r)+cap0(r)
*taxk0(r)+sum(j,-tm0(j,r))+sum((i,s,j),ijs0(i,j,s,r)
*taxe0(j,s))-cg0(r);
```

```
bop0(r)   =   sum(i,ex0(i,r)+im0(i,r));
trn0(r)   =   hsave0(r)+gsave0(r)-bop0(r)-inv0(r);
trn0("CheckSum") = sum(r,trn0(r));
display hsave0, gsave0, bop0, trn0;
set rnum(r)   Numeraire region;
rnum(r) = yes$(cp0(r) = smax(s,cp0(s)));
display rnum;

Parameter
  lab00(r)
  k00(r)
  gsave00(r)
  bop00(r)
  trn00(r);

lab00(r)    =   lab0(r);
k00(r)      =   k0(r);
gsave00(r)  =   gsave0(r);
bop00(r)    =   bop0(r);
trn00(r)    =   trn0(r);
```

```
*===============================================================
*    MPSGE model
*===============================================================
$ontext
$model:MRM_chn
```

```
$sectors:
  y(i,r)                    ! Production
  ya(i,r)                   ! Armington aggregate
  im(i,r)$im0(i,r)          ! Import
  ex(i,r)$ex0(i,r)          ! Export
  cp(r)                     ! Household consumption
  cg(r)                     ! Government capital consumption
  inv(r)                    ! Private capital formation

$commodities:
  pd(i,r)                   ! Price of output for domestic use
  pa(i,r)                   ! Price of armington aggregates
  pm(i,r)$im0(i,r)          ! Price of import
  px(i,r)$ex0(i,r)          ! Price of export
  pcp(r)                    ! Price of consumption
  pcg(r)                    ! Price of government consumption
  pinv(r)                   ! Price of investment
  pl(r)                     ! Wage rate
  rk(r)                     ! Capital service cost
  pfx                       ! Price of foreign exchange

$consumers:
  ha(r)                     ! Household agent
  gov(r)                    ! Government

$prod:y(j,r)$ex0(j,r)  t:2  s:0.1  va(s):1
  o:pd(j,r) q:yd0(j,r) p:py0(j,r) a:gov(r) t:taxy0(j,r)
  o:px(j,r) q:ex0(j,r) p:py0(j,r) a:gov(r) t:taxy0(j,r)
  i:pa(i,s) q:ijs0(i,j,s,r) P:pa0(j,r) a:gov(r) t:taxe0(j,r)
  i:pl(r)   q:labs0(j,r) p:pl0(r) a:gov(r) t:taxl0(r) p
  i:rk(r)   q:(caps0(j,r)/rk0(r)) p:rktx0(r) a:gov(r) t:taxk0(r) va:

$prod:y(j,r)$(not ex0(j,r))  s:0.1  va(s):1
```

```
    o:pd(j,r)   q:y0(j,r)  p:py0(j,r)  a:gov(r)  t:taxy0(j,r)
    i:pa(i,s)   q:ijs0(i,j,s,r) P:pa0(j,r) a:gov(r) t:taxe0(j,r)
    i:pl(r)     q:labs0(j,r) p:pl0(r)   a:gov(r)  t:taxl0(r)  va:
    i:rk(r)     q:(caps0(j,r)/rk0(r))  p:rktx0(r) a:gov(r) t:taxk0(r) va:

$prod:ya(j,r)    s:2
  o:pa(j,r)      q:ya0(j,r)
  i:pd(j,r)      q:yd0(j,r)
  i:pm(j,r)      q:(-im0(j,r)-tm0(j,r))

$prod:im(i,r)$im0(i,r)
  o:pm(i,r)      q:(-im0(i,r)-tm0(i,r))
  i:pfx q:(-im0(i,r))  p:pm0(i,r) a:gov(r) t:taxm0(i,r)

$prod:ex(i,r)$ex0(i,r)
  o:pfx          q:ex0(i,r)
  i:px(i,r)      q:ex0(i,r)

$prod:cp(r)    s:0.5
  o:pcp(r)       q:cp0(r)
  i:pa(i,s)      q:cps0(i,s,r)

$prod:cg(r)    s:0
  o:pcg(r)       q:cg0(r)
  i:pa(i,s)      q:cgs0(i,s,r)

$prod:inv(r)   s:0.5
  o:pinv(r)      q:inv0(r)
  i:pa(i,s)      q:invs0(i,s,r)
$demand:ha(r)
  d:pcp(r)       q:cp0(r)
  d:pinv(r)      q:inv0(r)
  e:pl(r)        q:lab0(r)
```

```
    e:rk(r)            q:k0(r)
    e:pcg(r)           q:gsave0(r)
    e:pfx              q:(-bop0(r))
    e:pcp(rnum)        q:(-trn0(r))
$demand:gov(r)
    d:pcg(r)           q:cg0(r)
    e:pcg(r)           q:(-gsave0(r))
$report:
    v:qd(i,r)          o:pd(i,r)       prod:y(i,r)
    v:qex(i,r)         o:px(i,r)       prod:y(i,r)
    v:ql(i,s,r)        i:pl(s)         prod:y(i,r)
    v:qk(i,s,r)        i:rk(s)         prod:y(i,r)
    v:qcp(r)           o:pcp(r)        prod:cp(r)
    v:qcg(r)           o:pcg(r)        prod:cg(r)
    v:qinv(r)          o:pinv(r)       prod:inv(r)
    v:qim(i,r)         o:pm(i,r)       prod:im(i,r)
$offtext
$sysinclude mpsgeset MRM_chn
Parameters
    rep1(*,*,*,*)
    rep_qd(*,*,*,*)
    rep_pd(*,*,*,*);
*===============================================================
*   Recursive solution:
*===============================================================
loop(t,
    y.l(i,r)    = qref(t,r);
    ya.l(i,r)$im0(i,r)  = qref(t,r);
    im.l(i,r)$ex0(i,r)  = qref(t,r);
    ex.l(i,r)   = qref(t,r);
    cp.l(r)     = qref(t,r);
    cg.l(r)     = qref(t,r);
    inv.l(r)    = qref(t,r);
    pd.l(i,r)   = pref(t,r);
```

```
   pa.l(i,r)    = pref(t,r);
   pm.l(i,r)$im0(i,r)  = pref(t,r);
   px.l(i,r)$ex0(i,r)  = pref(t,r);
   pcp.l(r)   = pref(t,r);
   pcg.l(r)   = pref(t,r);
   pinv.l(r)  = pref(t,r);
   pl.l(r)    = pref(t,r);
   rk.l(r)    = rk0(r)*pref(t,r);
   pfx.l  = sum(rnum,pref(t,rnum));
display hsave0, gsave0, bop0, trn0;
MRM_chn.iterlim = 1000;
MRM_chn.workspace = 24;
$include MRM_chn.gen
solve MRM_chn using mcp;
k0(r)   = (1-depr0)*k0(r)+qinv.l(r);
lab0(r)   =  lab0(r)*(1+gr0(t,r));
gsave0(r)  =  gsave0(r)*(1+gr0(t,r));
bop0(r)   =  bop0(r)*(1+gr0(t,r));
trn0(r)   =  trn0(r)*(1+gr0(t,r));
rep1("gdp",r,t,"base")   = qcp.l(r)+qcg.l(r)+qinv.l(r)+
sum(i,qex.l(i,r)-qim.l(i,r));
rep1("gdp","all",t,"base")   = sum(r,rep1("gdp",r,t,"base"));
rep1("g_gdp",r,t,"base")$(year(t) > 1997) = (rep1 ("gdp",
r,t,"base")/rep1("gdp",r,t-1,"base")-1)*100;
rep1("g_gdp","all",t,"base")$(year(t) > 1997)  = (rep1 ("gdp",
"all",t,"base")/rep1("gdp","all",t-1,"base")-1)*100;
rep1("cp",r,t,"base")    = qcp.l(r);
rep1("cg",r,t,"base")    = qcg.l(r);
rep1("inv",r,t,"base")   = qinv.l(r);
rep1("ex",r,t,"base")    = sum(i,qex.l(i,r));
rep1("im",r,t,"base")    = sum(i,qim.l(i,r));
rep1("pcp",r,t,"base")   = pcp.l(r);
rep1("g_pcp",r,t,"base")$(year(t) > 1997) = (rep1 ("pcp",
r,t,"base")/rep1("pcp",r,t-1,"base")-1)*100;
```

```
rep1("pinv",r,t,"base")  = pinv.l(r);
rep1("pl",r,t,"base")    = pl.l(r);
rep1("rk",r,t,"base")    = rk.l(r);
rep1("l0",r,t,"base")    = sum((i,s),ql.l(i,s,r));
rep1("k0",r,t,"base")    = sum((i,s),qk.l(i,s,r));
rep1("l1",r,t,"base")    = sum((i,s),ql.l(i,r,s));
rep1("k1",r,t,"base")    = sum((i,s),qk.l(i,r,s));
rep_qd(t,i,r,"base")     = qd.l(i,r);
rep_pd(t,i,r,"base")     = pd.l(i,r);

);
*display rep1;


*================================================================
*   Counter factural solution
*================================================================
k0(r)    = k00(r);
lab0(r)  = lab00(r);
gsave0(r) = gsave00(r);
bop0(r)  = bop00(r);
trn0(r)  = trn00(r);

loop(t,
  y.l(i,r)  = qref(t,r);
  ya.l(i,r) = qref(t,r);
  im.l(i,r)$im0(i,r)  = qref(t,r);
  ex.l(i,r)$ex0(i,r)  = qref(t,r);
  cp.l(r)   = qref(t,r);
  cg.l(r)   = qref(t,r);
  inv.l(r)  = qref(t,r);
  pd.l(i,r) = pref(t,r);
  pa.l(i,r) = pref(t,r);
  pm.l(i,r)$im0(i,r)  = pref(t,r);
  px.l(i,r)$ex0(i,r)  = pref(t,r);
```

```
  pcp.l(r)     = pref(t,r);
  pcg.l(r)     = pref(t,r);
  pinv.l(r)    = pref(t,r);
  pl.l(r)      = pref(t,r);
  rk.l(r)      = rk0(r)*pref(t,r);
  pfx.l   = sum(rnum,pref(t,rnum));
if(year(t)   = 1997,
 taxe0("cm","nowe")   = 0.05;
 taxe0("ogm","nowe")  = 0.05;
 taxe0("cm","sowe")   = 0.05;
 taxe0("ogm","sowe")  = 0.05;
 trn0(r)   = hsave0(r)+gsave0(r)-bop0(r)-inv0(r);
 trn0("CheckSum") = sum(r,trn0(r));

);

display hsave0, gsave0, bop0, trn0;
$set name gsave
$include MRM_chn.gen
solve MRM_chn using mcp;
k0(r)    = (1-depr0)*k0(r)+qinv.l(r);
lab0(r)  = lab0(r)*(1+gr0(t,r));
gsave0(r) = gsave0(r)*(1+gr0(t,r));
bop0(r)  = bop0(r)*(1+gr0(t,r));
trn0(r)  = trn0(r)*(1+gr0(t,r));
rep1("gdp",r,t,"alt")   =   qcp.l(r)+qcg.l(r)+qinv.l(r)+sum
(i,qex.l(i,r)-qim.l(i,r));
rep1("gdp","all",t,"alt") = sum(r,rep1("gdp",r,t,"alt"));
rep1("g_gdp",r,t,"alt")$(year(t) > 1997) = (rep1 ("gdp",
r,t,"alt")/rep1("gdp",r,t-1,"alt")-1)*100;
rep1("g_gdp","all",t,"alt")$(year(t) > 1997)  = (rep1 ("gdp",
"all",t,"alt")/rep1("gdp","all",t-1,"alt")-1)*100;
rep1("cp",r,t,"alt")  = qcp.l(r);
rep1("cg",r,t,"alt")  = qcg.l(r);
```

```
rep1("inv",r,t,"alt")   = qinv.l(r);
rep1("ex",r,t,"alt")    = sum(i,qex.l(i,r));
rep1("im",r,t,"alt")    = sum(i,qim.l(i,r));
rep1("pcp",r,t,"alt")   = pcp.l(r);
rep1("g_pcp",r,t,"alt")$(year(t) > 1997) = (rep1("pcp",r,
t,"alt")/rep1("pcp",r,t-1,"alt")-1)*100;
rep1("pinv",r,t,"alt")  = pinv.l(r);
rep1("pl",r,t,"alt")    = pl.l(r);
rep1("rk",r,t,"alt")    = rk.l(r);
rep1("l0",r,t,"alt")    = sum((i,s),ql.l(i,s,r));
rep1("k0",r,t,"alt")    = sum((i,s),qk.l(i,s,r));
rep1("l1",r,t,"alt")    = sum((i,s),ql.l(i,r,s));
rep1("k1",r,t,"alt")    = sum((i,s),qk.l(i,r,s));
rep_qd(t,i,r,"alt")     = qd.l(i,r);
rep_pd(t,i,r,"alt")     = pd.l(i,r);
);
rep1("gdp",r,t,"chg")   = (rep1("gdp",r,t,"alt")/rep1("gdp",
r,t,"base")-1)*100;
rep1("gdp","all",t,"chg") = (rep1("gdp","all",t,"alt")/rep1
("gdp","all",t,"base")-1)*100;
rep1("g_gdp",r,t,"dev") =  rep1("g_gdp",r,t,"alt")-rep1
("g_gdp",r,t,"base");
rep1("g_gdp","all",t,"dev") = rep1("g_gdp","all",t,"alt")
  -rep1("g_gdp","all",t,"base");
rep1("cp",r,t,"chg")    = (rep1("cp",r,t,"alt")/rep1("cp",r,t,
"base") -1)*100;
rep1("cg",r,t,"chg")    = (rep1("cg",r,t,"alt") /rep1("cg",
r,t,"base")-1)*100;
rep1("inv",r,t,"chg")   = (rep1("inv",r,t,"alt")/rep1("inv",
r,t,"base")-1)*100;
rep1("ex",r,t,"chg")    = (rep1("ex",r,t,"alt") /rep1("ex",
r,t,"base")-1)*100;
rep1("im",r,t,"chg")    = (rep1("im",r,t,"alt") /rep1("im",
r,t,"base")-1)*100;
```

```
rep1("pcp",r,t,"chg")   =   (rep1("pcp",r,t,"alt")/rep1("pcp",
r,t,"base")-1)*100;
rep1("g_pcp",r,t,"chg") =   rep1("g_pcp",r,t,"alt")-rep1("g_pcp",
r,t,"base");
rep1("pl",r,t,"chg")    =   (rep1("pl",r,t,"alt") /rep1("pl",
r,t,"base")-1)*100;
rep1("rk",r,t,"chg")    =   (rep1("rk",r,t,"alt") /rep1("rk",
r,t,"base")-1)*100;
rep1("l0",r,t,"chg")    =   (rep1("l0",r,t,"alt") /rep1("l0",
r,t,"base")-1)*100;
rep1("k0",r,t,"chg")    =   (rep1("k0",r,t,"alt") /rep1("k0",
r,t,"base")-1)*100;
rep1("l1",r,t,"chg")    =   (rep1("l1",r,t,"alt") /rep1("l1",
r,t,"base")-1)*100;
rep1("k1",r,t,"chg")    =   (rep1("k1",r,t,"alt") /rep1("k1",
r,t,"base")-1)*100;
rep_qd(t,i,r,"chg")     =   (rep_qd(t,i,r,"alt")/rep_qd(t,i,r,
"base")-1)*100;
rep_pd(t,i,r,"chg")     =   (rep_pd(t,i,r,"alt")/rep_pd(t,i,r,
"base")-1)*100;
display rep1;
Execute_Unload 'chn_dyn_%name%.gdx', rep1,rep_qd, rep_pd;
```

Acknowledgment

This study was financially supported by China Scholarship Council and Tohoku University Ecosystem Adaptability Global COE Program. The publication of this book is supported by "the Fundamental Research Funds for the Central Universities" (Fund No.: 3214007103) and "Jiangsu Research Base of Innovation Driven Development". I gratefully acknowledge the help of my supervisor Professor Hayashiyama Yasuhisa for all the guidance he gave and all the time he spent. Also, I would like to thank Mr. Abe Masahiro and Ms. Yan Rong for their help on my research and this book.